Optical Radiation Detectors

Optical Radiation Detectors

EUSTACE L. DERENIAK

Associate Professor
Optical Sciences Center
University of Arizona

DEVON G. CROWE

Senior Scientist and Manager
Advanced Concepts Branch
Sensor Technology Division
Science Applications, Inc.

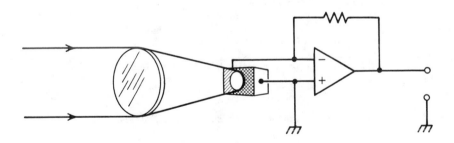

John Wiley & Sons
New York / Chichester / Brisbane / Toronto / Singapore

Library of Congress Cataloging in Publication Data:

Dereniak, Eustace L.
    Optical radiation detectors.

    (Wiley series in pure & applied optics, ISSN 0277-2493)
    Includes index.
    1. Radiation—Instruments. 2. Radiation—Measurement.
3. Optical detectors. I. Crowe, Devon G. II. Title.
III. Series: Wiley series in pure and applied optics.

QC338.D47   1984   539.7'7   84-7356
ISBN 0-471-89797-3

Printed in the United States of America

10 9 8 7 6 5 4

To our
wives and families

Preface

The application of optical detectors to solve myriads of technical problems in medicine, industry, and basic research has generated a large demand for scientists and engineers who are knowledgeable in the subject. This book is intended to fill the need for a textbook that will allow the college student to become a productive user of detector technology. It is hoped that the reader who desires to learn about detectors will find a coherent presentation from introductory theory through the practical application of detector technology.

The material contained in this book has been developed from a one-semester graduate course that has been taught at the Optical Sciences Center of the University of Arizona for the past 5 years. The level of presentation is suitable for first-year graduate students in physics and engineering who have received the ordinary preparation in modern physics and elementary electronic ciruits. The treatment includes introductory theory of operation for students with no prior work in the field. The theoretical treatment is sufficiently complete for the reader to calculate the fundamental limits to detector performance, as well as to calculate the predicted signal-to-noise ratio of an overall system in which the detector is a part. In addition, each chapter has a number of selected problems covering the material of that particular chapter. These problems have been student-tested both in coursework and in preliminary examinations for the Ph.D. degree.

We wish to acknowledge the assistance of the following individuals during the preparation of this book: William J. White of Honeywell, Inc., Lexington, MA; Richard W. Capps of the Institute for Astronomy, University of Hawaii; Chris Perry of Aerojet Electrosystems Inc., Azusa, CA; Lee Hudson of the Naval Weapons Center, China Lake, CA; Pat Britt of the Kitt Peak National Observatory, Tucson, AZ; Steven Dziuban of Rockwell; Fred E. Nicodemus of the National Bureau of Standards; Fred Bartell of the University of Alabama at Huntsville; Lang Brod, Scott Devore, Carol Hickey, Barbara Landesman, Li Lifeng, Steve Martinek, Phyllis Miller, Allan Craig, Keith Prettyjohns, and Dave Perry of the

Optical Sciences Center of the University of Arizona; and Molly Chambers, Douglas J. Granrath, Lucia J. Homicki, and Karen McCaffrey of Science Applications, Inc., Tucson, AZ. Responsibility for any remaining errors in this book must, of course, reside with us.

We wish to extend a special acknowledgment to Professor William L. Wolfe of the Optical Sciences Center of the University of Arizona for serving as teacher, advisor, and source of ideas throughout the graduate schooling and subsequent careers of both of us.

Finally, we wish to express our gratitude to Beatrice Shube, Christina Mikulak, Rosalyn Farkas, and the Wiley production staff for their excellent support in the preparation of the book.

EUSTACE L. DERENIAK
DEVON G. CROWE

Tucson, Arizona
June 1984

Contents

Optical Radiation Detectors

CHAPTER 1

RADIOMETRY

1-1 INTRODUCTION

Radiometry is the science of the measurement of electromagnetic radiation. This chapter reviews radiometric concepts that are prerequisites to the analysis of detector performance. Physical units, thermal (blackbody) radiation, radiative energy transfer, and "photon noise" or "quantum noise" (the uncertainty in the amount of energy transferred) must all be understood in order to appreciate the significance of the figures of merit of detector performance that are introduced in later chapters. These concepts are also used in the calculation of the predicted performance of optical systems that utilize detectors. For a more detailed consideration of radiometry, the reader should consult the standard reference works by Nicodemus (1976, 1978); Wolfe and Zissis (1978); and Grum and Becherer (1979).

The nomenclature and symbology for the radiometric quantities used in this book are given in Table 1-1. This is similar to Nicodemus (1976). This book will not use photometric quantities, which are based on the approximate spectral response of the human eye. Appendix A gives a brief introduction to photometric quantities for reference purposes. Note that the symbols for each radiometric quantity have either the subscript e for energy-derived units or p for photon-derived units.

1-2 THERMAL RADIATION

All real objects in the universe are thought to exist at temperatures above absolute zero, which means that the atoms and molecules that compose the

1

Table 1-1
Nomenclature for Radiometric Quantities

Quantity	Symbol	Units	Unit Symbol
Radiant energy	Q_e	Joule	J
Radiant flux (power)	ϕ_e	Watt	
Radiant intensity	I_e	Watt per steradian	$W\,sr^{-1}$
Radiant-flux density			
Radiant exitance	M_e	Watt per square centimeter	$W\,cm^{-2}$
Irradiance	E_e	Watt per square centimeter	$W\,cm^{-2}$
Radiance	L_e	Watt per steradian and square centimeter	$W\,sr^{-1}\,cm^{-2}$
Photon energy	Q_p	Photon	q^a
Photon flux	ϕ_p	Photon per second	$q\,sec^{-1}$
Photon flux intensity	I_p	Photon per second and steradian	$q\,sec^{-1}\,sr^{-1}$
Photon flux exitance	M_p	Photon per second and square centimeter	$q\,sec^{-1}\,cm^{-2}$
Incident-photon-flux density	E_p	Photon per second and square centimeter	$q\,sec^{-1}\,cm^{-2}$
Photon flux radiance (sterance)	L_p	Photon per second and steradian and square centimeter	$q\,sec^{-1}\,sr^{-1}\,cm^{-2}$

^{a}q = quanta.

2

objects are in motion. These motions are constrained by interactions with other atoms and molecules (e.g., collisons and bonds); therefore, the elementary charges within these atoms are subjected to accelerations. The accelerating charges radiate electromagnetically.

An object that is a perfect emitter ("blackbody") of thermal electromagnetic radiation follows the Planck law:

3.7418

$$M_{e,\lambda}(\lambda, T) = \frac{2\pi hc^2}{\lambda^5[e^{hc/\lambda kT} - 1]} = \frac{3.74 \times 10^4}{\lambda^5[e^{14,388/\lambda T} - 1]} \tag{1.1}$$

where $M_{e,\lambda}(\lambda, T)$ = spectral radiant exitance in watts per square centimeter of area and micrometer of radiation wavelength (W cm^{-2} μm^{-1}),

λ = emitted wavelength in micrometers (μm),

T = absolute temperature of the blackbody in kelvins (K),

h = Planck's constant (6.626176×10^{-34} W sec^2),

c = speed of light ($2.99792438 \times 10^{10}$ cm sec^{-1}),

k = Boltzmann's constant (1.380662×10^{-23} W sec K^{-1}).

Using the expression for photon energy,

$$Q_p = \frac{Q_e \lambda}{hc} \tag{1.2}$$

one may convert Eq. (1.1) into an expression for the spectral photon flux exitance:

1.88365

$$M_{p,\lambda}(\lambda, T) = \frac{2\pi c}{\lambda^4[e^{hc/\lambda kT} - 1]} = \frac{1.885 \times 10^{23}}{\lambda^4[e^{14,388/\lambda T} - 1]} \tag{1.3}$$

where $M_{p,\lambda}(\lambda, T)$ has units of photons per second and square centimeter and micrometer (q sec^{-1} cm^{-2} μm^{-1}; λ in micrometers, T in kelvins).

The total emission in all wavelengths can be obtained by integrating Eqs. (1.1) and (1.3) over all wavelengths. The total energy expression is known as the Stefan–Boltzmann law:

$$M_e(T) = \int_0^\infty M_{e,\lambda}(\lambda, T)d\lambda = \frac{2\pi^5 k^4 T^4}{15h^3c^2} = \sigma_e T^4 \tag{1.4}$$

where $\sigma_e = 5.67032 \times 10^{-12}$ W cm^{-2} K^{-4}. The corresponding photon flux exitance expression is

$$M_p(T) = \int_0^\infty M_{p,\lambda}(\lambda, T)d\lambda = \sigma_p T^3$$

$$\sigma_e \cong 5.67 \times 10^{-12} \frac{watt}{K^4 cm^2}$$

where

$$\sigma_p = 1.52 \times 10^{11} \text{ photons sec}^{-1} \text{ cm}^{-2} \text{ K}^{-3} \tag{1.5}$$

While a visiting lecturer at the Institute of Electro-Optical Engineering of the National Chaio Tong University in Taiwan, one of the authors (Dereniak) was hosted by a scholar named Uen Tzeng-Ming who need only introduce himself in Chinese to summarize these results. His name translates as "Temperature Increases Brightness."

Equations (1.4) and (1.5) indicate that at room temperature (~300 K), a 1 cm² blackbody emits about 50 mW over all wavelengths in the form of about 4×10^{18} photons sec^{-1}. However, nearly all of these photons have a wavelength that is too long for the eye to respond to, so that the thermal emission of room-temperature objects can only be imaged by using infrared detection equipment.

Wien's displacement law gives a relation between temperature and the wavelength peak of the Planck function [Eq. (1.1)]. This relation is obtained by setting the wavelength derivative of Eq. (1.1) to zero,

$$\frac{\partial M_{e,\lambda}(\lambda,\,T)}{\partial\lambda} = 0 \tag{1.6}$$

and solving for the value of λT which maximizes $M_{e,\lambda}(\lambda,\,T)$:

$$\lambda_{\max} T = 2898 \ \mu\text{m K} \tag{1.7}$$

This equation is valid for spectral radiant exitance in watts per square centimeter and micrometer. A similar process using Eq. (1.3) for $M_{p,\lambda}(\lambda,\,T)$ photons per second per square centimeter and micrometer yields the peak of $M_{p,\lambda}(\lambda,\,T)$:

$$\lambda_{\max} T = 3669 \ \mu\text{m K} \tag{1.8}$$

The fact that Eqs. (1.7) and (1.8) yield different values of $\lambda_{\max}$ should not be interpreted to mean that photon detectors may have an advantage or disadvantage in detecting a source of a given temperature at a given wavelength when compared to an energy detector. In fact, the peak of a monochromatic function should not be interpreted as carrying any precise physical meaning. It is only the integrals over finite wavelength intervals that can be physically significant. If Eqs. (1.1) and (1.3) are expressed in terms of frequency, ν (in hertz), instead of wavelengths, we have

$$M_{e,\nu}(\nu,\,T) = \frac{2\pi h\nu^3}{c^2[e^{h\nu/kT} - 1]} \tag{1.9}$$

and

$$M_{p,\nu}(\nu, T) = \frac{2\pi\nu^2}{c^2[e^{h\nu/kT} - 1]}$$ (1.10)

The frequencies at which Eqs. (1.9) and (1.10) peak do not correspond to the wavelengths at which Eqs. (1.1) and (1.3) peak:

$$\nu_{max} \neq \frac{c}{\lambda_{max}}$$ (1.11)

In fact, $M_{e,\nu}(\nu, T)$ peaks at

$$\frac{\nu_{max}}{T} = 5.876 \times 10^{10} \text{ Hz/K}$$ (1.12)

from which one obtains a ν_{max} that corresponds to approximately $1.76\lambda_{max}$, where λ_{max} is calculated from the Wien displacement law. All real physical systems observe thermal radiation with a finite spectral resolution $d\lambda$ or $d\nu$, where

$$d\nu = \frac{-c \, d\lambda}{\lambda^2}$$ (1.13)

The minus sign in this relation arises because ν decreases as λ increases. Equation (1.13) ensures that Eqs. (1.1) and (1.9) give the correct (same) answer when integrated over a finite interval.

The concept of emissivity is introduced to quantify the different emission properties of different objects. The spectral emissivity is defined to be the ratio of the radiance of the object under consideration to that of a perfect blackbody:

$$\varepsilon(\theta, \phi, \lambda) = \frac{L_{e,\lambda}(\lambda, T)_{\text{actual}}}{L_{e,\lambda}(\lambda, T)_{\text{blackbody}}}$$ (1.14)

where θ and ϕ define the direction. Note that $0 \leq \varepsilon(\theta, \phi, \lambda) \leq 1$. The total emissivity over the entire spectrum is then

$$\varepsilon = \frac{\int_0^\infty \varepsilon_e(\lambda) L_{e,\lambda}(\lambda, T) d\lambda}{\int_0^\infty L_{e,\lambda}(\lambda, T) d\lambda}$$ (1.15)

It is common to omit the temperature dependence of ε from notation, but the reader should remember the obvious temperature dependence of the definition. It is also common to assume that the energy emissivity is implied when the context makes it obvious that the photon emissivity is not being used. The photon total emissivity, on the other hand, is

$$\varepsilon = \frac{\displaystyle\int_0^\infty \varepsilon_p(\lambda) L_{p,\lambda}(\lambda, T) d\lambda}{\displaystyle\int_0^\infty L_{p,\lambda}(\lambda, T) d\lambda} \tag{1.16}$$

The photon emissivity of Eq. (1.16) is equal to the energy emissivity of Eq. (1.15) if $\varepsilon_e(\lambda) = \varepsilon_p(\lambda) = $ constant for all λ; otherwise, the two emissivities are, in general, different. The spectral emissivities will be the same for all cases, however:

$$\varepsilon_e(\lambda) = \frac{L_{e,\lambda}(\lambda, T)_{\text{object}}}{L_{e,\lambda}(\lambda, T)_{\text{blackbody}}} = \frac{(\lambda/hc) L_{p,\lambda}(\lambda, T)_{\text{object}}}{(\lambda/hc) L_{p,\lambda}(\lambda, T)_{\text{blackbody}}} = \varepsilon_p(\lambda) \tag{1.17}$$

For a perfect thermal radiator (blackbody), $\varepsilon(\lambda) \equiv 1$ for all λ.

Kirchhoff's law for electromagnetic radiation states that in thermodynamic equilibrium the amount of energy absorbed is equal to the amount of energy emitted. Therefore, for all surfaces,

$$\alpha(\lambda) = \varepsilon(\lambda) \tag{1.18}$$

where $\alpha(\lambda)$ is the absorptance or the ratio of the amount of radiation absorbed to the amount of radiation incident monochromatically. A blackbody, therefore, has $\varepsilon(\lambda) = 1$ and absorbs all of the incident radiation. The origin of the name blackbody is now evident: If a body absorbs all incident radiation, it can reflect no light and, hence, appears black.

There are only three possible components to the behavior of light incident on a surface: reflectance, transmittance, and absorptance (Grum and Becherer, 1979):

$$\alpha + \rho + \tau = 1 \tag{1.19}$$

where $\rho = $ reflectance or fraction of incident power reflected,

$\tau = $ transmittance (commonly called "transmission") or the fraction of incident power that emerges from the other side of the surface.

For a blackbody, $\alpha = 1$, so $\rho = \tau = 0$.

1-3   SOLID ANGLE

Radiometry (in this volume) is concerned with the radiative energy transfer between surfaces through three-dimensional Euclidean space. Therefore, the concept of solid angle is often useful in radiometric calculations.

The common definition of solid angle is the three-dimensional angular spread at the vertex of a cone measured by the area intercepted by the cone on a unit sphere (Fig. 1-1). This definition gives rise to the convenient unit of the steradian for measuring solid angle. The subtended solid angle in steradians is the ratio of the area of the spherical surface intercepted by the cone to the square of the radius of the sphere. The total area of a spherical surface is $4\pi r^2$, so a spherical surface subtends 4π steradians (sr).

In order to calculate the subtense of the incremental solid angle we note that

$$dA_s = r\, d\theta\, r \sin\theta\, d\phi = r^2 \sin\theta\, d\theta\, d\phi \qquad (1.20)$$

The incremental solid angle $(d\Omega)$ in steradians is

$$d\Omega = \frac{dA_s}{r^2} = \sin\theta\, d\theta\, d\phi \qquad (1.21)$$

Therefore, the measure of a general solid angle Ω is

$$\Omega = \int_\theta \int_\phi \sin\theta\, d\theta\, d\phi \qquad (1.22)$$

For a spherical surface intersected by a right circular cone symmetric about the z axis,

$\angle$ ½ angle $(\leq \pi/2)$

$$\Omega = \int_0^\theta \int_0^{2\pi} \sin\theta\, d\theta\, d\phi = 2\pi[1 - \cos\theta] \qquad (1.23)$$

Equation (1.23) yields a simple formula for calculating the number of steradians subtended by a spherical surface section with circular symmetry when the associated cone subtends a plane angle of 2θ $(0 \leq \theta \leq \pi/2)$. It is common practice to use the steradian unit to measure the angular subtense of any surface (spherical or not) of area A at distance r by using

$$\Omega \approx \frac{A}{r^2} \qquad (1.24)$$

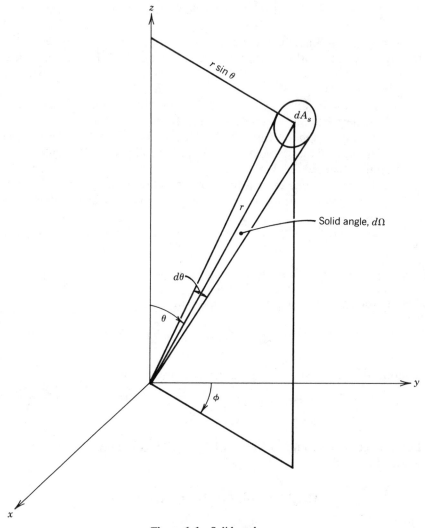

Figure 1-1 Solid angle.

Obviously, the smaller the angle ($r^2 \gg A$) becomes, and the closer the surface area A is to that of a spherical section, the more closely approximation (1.24) approaches Ω. For most practical problems, r^2 is in fact much larger than A, and this practice is justified.

1-4   RADIATIVE ENERGY TRANSFER

It is necessary, in general, to take into account the geometry of the source–detector system to calculate the signal incident on the detector. If the source is sufficiently well collimated that all of the emitted radiation falls on the detector, then only the power emitted by the source need be known. The usual case is that the detector only intercepts a small fraction of the radiated signal, however. This section summarizes some results that are used in the remaining portion of the book to calculate the signal incident on the detector.

Assuming that even a rough surface (i.e., a surface with irregularities that are large compared to the radiation wavelength) can be treated as a collection of plane surfaces, it is evident from Fig. 1-2 that the incremental power incident on the portion of the detector considered (dA_d) due to the portion of the source under consideration (dA_s) is given by

$$d\Phi = \frac{L^s dA_s \cos \theta_s dA_d \cos \theta_d}{R^2} \tag{1.25}$$

For the case of a plane source normal to the line of sight to a plane detector, which is also normal to the optical axis,

$$\Phi = \frac{L^s A_s A_d}{R^2} \tag{1.26}$$

Φ and L should both be expressed in terms of energy or photon flux using the subscript e or p, respectively. In most cases, $A_d \ll R^2$, so that

$$\Phi = L^s A_s \Omega_d \tag{1.27}$$

since $\Omega_d \approx A_d/R^2$ is the solid angle subtended by the detector from the source. It is also often true that

$$\Phi = L^s A_d \Omega_s \tag{1.28}$$

This can be thought of as a consequence of the $A\Omega$ product being constant. The product $A_d \Omega_s$ is known as the optical throughput. If the

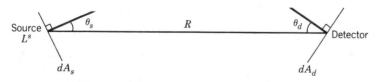

Figure 1-2   Source–detector system geometry.

field of view of the detector system is greater than the solid angle subtended by the source (Ω_s) at the detector, a large uniform background of radiance L^B contributes a radiant power of

$$\Phi = L^B A_d (\Omega_B - \Omega_S) \tag{1.29}$$

where $\Omega_B > \Omega_S$ and Ω_B represents the solid angle field of view intercepting the background.

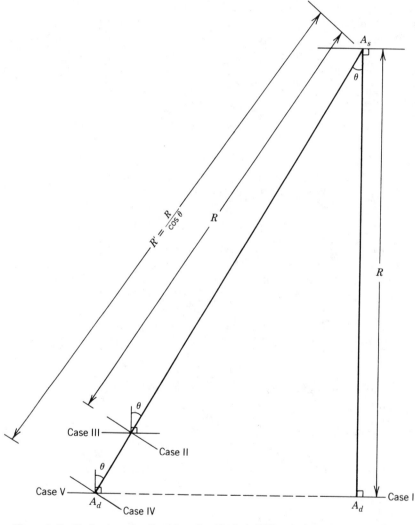

Figure 1-3 Cosine law of radiometry: $\Phi = (L^s A_s A_d / R^2) \cos^{n-1} \theta$; $n = $ case number.

A derivation of the cosine law of radiometry is illustrated in Fig. 1-3, which describes geometry configurations of Cases I, II, III, IV, and V. Case I has no cosine dependence; Case II has cosine dependence; Case III has $(\text{cosine})^2$ dependence; Case IV has $(\text{cosine})^3$ dependence; and Case V has $(\text{cosine})^4$ dependence (Palmer, 1978). The cosine factors accumulate to include area projection and distance-increase effects. It is possible to reach Case V in two steps. First, the expression for Case III is derived from Eq. (1.25):

$$\Phi = \frac{L^s A_s \cos\theta_s A_d \cos\theta_d}{R^2} = \frac{L^s A_s A_d}{R^2}\cos^2\theta \tag{1.30}$$

Then we note that

$$R'^2 = \left(\frac{R}{\cos\theta}\right)^2 = \frac{R^2}{\cos^2\theta} \tag{1.31}$$

For Case V we should use R' instead of R in Eq. (1.30) and that substitution results in

$$\Phi = \frac{L^s A_s A_d}{R'^2}\cos^2\theta = \frac{L^s A_s A_d}{R^2/\cos^2\theta}\cos^2\theta = \frac{L^s A_s A_d}{R^2}\cos^4\theta \tag{1.32}$$

From Eq. (1.32) it is evident that the signal varies across the detector plane as the fourth power of the cosine of the angle off-axis.

The cosine law of radiometry implies the often quoted "cosine-to-the fourth law" of detector plane irradiance. The lens in Fig. 1-4 is assumed to be free of aberrations and to transmit all the incident radiation to the

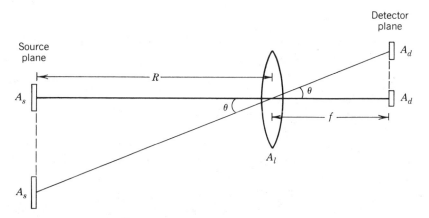

Figure 1-4   Detector plane irradiance.

detector plane ($\tau = 1$). The power on the detector is then given by Eq. (1.32) if A_d is replaced by A_l, the collecting lens area. The average irradiance across the detector itself is

$$E = \frac{\Phi}{A_d} \qquad (1.33)$$

Therefore,

$$E = \frac{L^s A_s A_l}{A_d R^2} \cos^4 \theta \qquad (1.34)$$

We now consider A_s to denote only the portion of an extended source that is imaged by the lens onto the detector area A_d. Then, we have that the solid angle subtended by the detector at the lens is the same as the solid angle subtended by A_s at the lens:

$$\Omega = \frac{A_d}{f^2} = \frac{A_s}{R^2} \qquad (1.35)$$

Using Eq. (1.35), we modify Eq. (1.34):

$$E = \frac{L^s A_d A_l}{A_d f^2} \cos^4 \theta = \frac{L^s A_l}{f^2} \cos^4 \theta \qquad (1.36)$$

This equation is commonly called the "$\cos^4$ law." In fact, with real lens aberrations taken into account (primarily coma), one finds experimentally that for a given optical system

$$E \approx \frac{L^s A_l}{f^2} \cos^n \theta \qquad (1.37)$$

where the value of n is usually between 2.5 and 4 depending on the field size and aberration corrections of the lens.

1-5 LAMBERTIAN RADIATORS

The concept of a Lambertian radiator is introduced because a mathematical analysis of its radiometric properties is instructive and relatively simple and because many important sources are approximately Lambertian. A Lambertian surface is one for which the surface radiance (W sr^{-1} cm^{-2}) is independent of the angle from which it is viewed. Alternately, a Lambertian radiator is an isotropically diffuse surface for

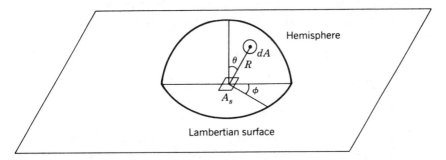

Figure 1-5 Lambertian source.

which the radiant intensity (I) in any direction varies as the cosine of the angle between that direction and the normal to the surface. The cosine term arises from the angular dependence of the projected area. Reiterating, the radiance is the same in all directions from surface.

The directional properties of a Lambertian source can be calculated using the terms illustrated in Fig. 1-5. The power incident on a surface element of the hemisphere is

$$d\Phi = \frac{L^s dA_s \cos\theta \, dA}{R^2}$$

$$= \frac{L^s dA_s \cos\theta \, R \sin\theta \, d\theta \, R \, d\phi}{R^2}$$

$$= L^s dA_s \cos\theta \sin\theta \, d\theta \, d\phi \qquad (1.38)$$

The total radiant exitance can now be found:

$$M = \int_{\text{hemisphere}} \frac{d\phi}{dA_s} = L^s \int_0^{\pi/2} \int_0^{2\pi} \cos\theta \sin\theta \, d\theta \, d\phi \qquad (1.39)$$

Integrating,

$$M = \pi L^s \qquad (1.40)$$

for a Lambertian source. Equation (1.40) is sometimes interpreted as "a Lambertian source radiates into π steradians." The quoted statement is of course incorrect—the form of Eq. (1.40) is due to the cosine dependence of areal projection throughout 2π steradians.

The effect of the field of view of the detector system on the irradiance due to an extended uniform background is mathematically similar to the calculation (1.39), except that it is carried out over a symmetrical field of view of half-angle $\theta_{1/2}$ rather than over a full hemisphere:

$$E = L^s \int_0^{\theta_{1/2}} \int_0^{2\pi} \cos \theta \sin \theta \, d\theta \, d\phi \qquad (1.41)$$

Integrating,

$$E = \pi L^s \sin^2 \theta_{1/2} \qquad (1.42)$$

Consider the two cases (A and B) of a detector directly irradiated by (A) an extended source with field of view restricted by a circular stop and (B) a detector irradiated by a concentrating lens that has $\tau = 1$ and no aberrations. If the apparent fields of view of the detectors are the same (i.e., $\theta_{1/2}$ is the same in both cases), then the irradiance at the detector plane is the same in both cases (Fig. 1-6).

The irradiance cannot be increased by using a lens system. The irradiance can be decreased, however, if the lens has aberrations that spread the power received from a small source over a larger area or if the lens reflects or absorbs some of the incident radiation ($\tau < 1$). The power received from a point (unresolved) source is spread by diffraction and the irradiance is a function of the lens diameter (D) because of the functional dependence of the area of the Airy disc:

$$A_{\text{image}} = \pi \left(1.22 \frac{\lambda f}{D} \right)^2 \qquad (1.43)$$

As the lens diameter increases, the image area decreases, and the irradiance increases until the object is resolved. This simplified discussion has ignored atmospheric turbulence, which may cause the irradiance to remain constant with increasing aperture after reaching a system resolution equivalent to the "seeing disc." In every case, the irradiance at the image plane is less than or equal to the ideal situation of Eq. (1.42).

1-6   RADIATION NOISE

There is uncertainty in the amount of electromagnetic radiation emitted by any signal source within a fixed period of time. The statistical nature of signal sources will be introduced in this section.

A blackbody at 300 K exhibits a mean photon exitance flux density of about 4×10^{18} q sec^{-1} cm^{-2}. The actual number of photons emitted by a 1 cm^2 blackbody at 300 K during any specific 1-sec interval may vary considerably from 4×10^{18}. We merely expect an average of 4×10^{18} if

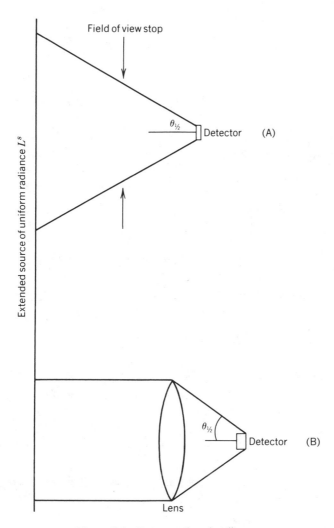

Figure 1-6   Detector plane irradiance.

many 1-sec intervals are sampled. This variability is illustrated in Fig. 1-7, which shows a most probable (or expected) value $\bar{M}_p$. This value is called the average or mean number of photons emitted from 1 cm² in 1 sec. The central-limit theorem can generally be expected to apply; therefore, for large numbers of photons per sample, a Gaussian distribution provides a good description of the variability. The standard deviation (σ) is a measure of the uncertainty or "noise" present in the photon emission rate.

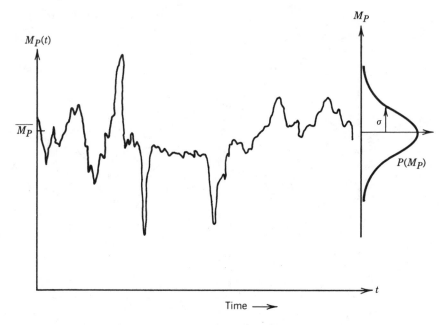

Figure 1-7   Variability of instantaneous photon exitance.

The probability of n photons being emitted in a period of time T can often be accurately estimated using the Poisson probability density function (Papoulis, 1968):

$$P_T(n) = \frac{\bar{n}^n}{n!e^{\bar{n}}} \qquad (1.44)$$

where $\bar{n}$ is the average number of photons emitted in a period of length T for a large number of samples (k):

$$\bar{n} = \frac{\displaystyle\sum_{i=1}^{k} n_i}{k} \qquad (1.45)$$

The Poisson distribution is used because it yields good agreement with experimental estimates of the distribution of any discrete physical process, such as the emission of particles from a radioactive source or the number of raindrops in a constant downpour. A discrete process is one in which a particle, raindrop, photon, or other discrete entity can be determined to be present or not present. This binary nature of each individual event in the process leads one to expect a binomial distribution of the events (provided that the events are random and time independent):

$$P(m) = \binom{n}{m} p^m (1-p)^{n-m} \tag{1.46}$$

where $P(m)$ is the probability of m events in n observations, p is the probability of one event in one observation, and

$$\binom{n}{m} = \frac{n!}{m!(n-m)!} \tag{1.47}$$

As the number of observations n becomes large, the binomial distribution approaches the Poisson distribution, until the two distributions become identical in the limit as $n \to \infty$. It will be assumed that the photon-emission process follows a Poisson distribution even though it is not possible to directly test this assumption. It is true that if the photon-emission process is Poisson, then a photon detector with a quantum collection efficiency of less than one will detect the arrival of m photons in a period T with a probability given by a Poisson density function:

$$P_T(m) = \frac{(\eta \bar{n})^m}{m! e^{(\eta \bar{n})}} \tag{1.48}$$

where $\bar{n}$ is the average number of photons incident on the detector in a time period T, and η is the detector quantum efficiency, or fraction of incident photons detected:

$$\eta = \frac{\bar{m}}{\bar{n}} \tag{1.49}$$

It cannot necessarily be inferred, however, that because the *detected* photon stream is described by Eq. (1.48), the *emitted* photon stream must be described by Eq. (1.44). The distribution of detected photons $[P_T(m)]$ can be derived from a Bernoulli distribution only if the emission distribution $[P_T(n)]$ is known:

$$P_T(m) = \sum_{n=m}^{\infty} P_T(n) \binom{n}{m} \eta^m (1-\eta)^{n-m} \tag{1.50}$$

but, in general, $P_T(m)$ has a different mathematical form than $P_T(n)$.

The fact that there is noise (uncertainty in magnitude) present in the signal itself implies a limit to the signal-to-noise ratio attainable with even a perfect, noiseless detector. It is well known that the variance of a Poisson distribution is equal to the mean:

$$\sigma^2_{\text{var}} = \bar{n} \qquad (1.51)$$

The root mean square (rms) noise level is defined to be the square root of the variance:

$$\sigma = \sqrt{\bar{n}} \qquad (1.52)$$

The signal-to-noise ratio present in the incident signal itself is therefore

$$\text{SNR} = \frac{\bar{n}}{\sigma} = \frac{\bar{n}}{\sqrt{\bar{n}}} = \sqrt{\bar{n}} \qquad (1.53)$$

For the detector with less than perfect quantum efficiency, but which is itself noiseless,

$$\text{SNR} = \frac{\eta\bar{n}}{\sqrt{\eta\bar{n}}} = \sqrt{\eta\bar{n}} \qquad (1.54)$$

The *observed* signal-to-noise ratio can never exceed that given by Eq. (1.54). This limit is only achieved by a noiseless photon detection system that is irradiated by a Poisson source.

Many optical-radiation sources do not meet the assumption of time independence for photon-emission events, which leads to the Poisson distribution. In fact, only coherent light is strictly Poisson, because only perfectly coherent light is truly independent of time. A blackbody source, for example, is incoherent or "chaotic" with respect to time. It may be instructive for the reader to view the result of Eq. (1.52) as the lower limit to the noise increased by time-dependent fluctuations that add to the uncertainty:

$$\sigma = \sqrt{\bar{n}} + \sigma_t \qquad (1.55)$$

where σ_t is the time-dependent component. In order to calculate the amount of noise present in a blackbody signal, however, we must use Bose–Einstein statistics.

The Bose–Einstein expression for the probability that the total photon energy at a specified frequency ν has a value u is

$$P(u) = \frac{e^{-u/kT}}{\displaystyle\int_0^\infty e^{-u/kT}\,du} \qquad (1.56)$$

The expected energy level (most probable value of u) is

$$\bar{u} = \frac{\int_0^\infty u e^{-u/kT} \, du}{\int_0^\infty e^{-u/kT} \, du} = \frac{h\nu e^{-h\nu/kT}}{1 - e^{-h\nu/kT}} \qquad (1.57)$$

The expected number of photons, each of energy $h\nu$, must be

$$\bar{n} = \frac{\bar{u}}{h\nu} = \frac{e^{-h\nu/kT}}{1 - e^{-h\nu/kT}} \qquad (1.58)$$

Similarly, the expected value of n^2 can be found:

$$\overline{n^2} = \frac{\overline{u^2}}{(h\nu)^2} = \frac{(1/h^2\nu^2)\int_0^\infty u^2 e^{-u/kT} \, du}{\int_0^\infty e^{-u/kT} \, du} = \frac{e^{-h\nu/kT} + e^{-2h\nu/kT}}{(1 - e^{-h\nu/kT})^2} \qquad (1.59)$$

The variance can now be found:

$$\sigma^2 = \overline{n^2} - \bar{n}^2 = \frac{e^{-h\nu/kT} + e^{-2h\nu/kT}}{(1 - e^{-h\nu/kT})^2} - \frac{e^{-2h\nu/kT}}{(1 - e^{-h\nu/kT})^2}$$

$$= \frac{e^{-h\nu/kT}}{(1 - e^{-h\nu/kT})^2} \qquad (1.60)$$

Note that Eq. (1.60) can be rewritten:

$$\sigma^2 = \bar{n}\left[\frac{e^{h\nu/kT}}{e^{h\nu/kT} - 1}\right] \qquad (1.61)$$

which is the variance for a Poisson distribution ($\sigma^2 = \bar{n}$) multiplied by the Boson factor. For the majority of optical applications, we are interested in wavelengths from 0.3 to 30 μm and in $T < 500$ K. These values imply that the energy per photon is much greater than the thermal energy ($h\nu \gg kT$), and, therefore,

$$\frac{e^{h\nu/kT}}{e^{h\nu/kT} - 1} \approx 1 \qquad (1.62)$$

and we can use the Poisson result for photon noise ($\sigma^2 = \bar{n}$). When very-high-temperature sources are to be detected, or when working at long wavelengths approaching a millimeter or more, the correction factor to Eq. (1.51) should be remembered and applied.

BIBLIOGRAPHY

Bartell, F. O. and W. L. Wolfe, "Cavity radiators: an ecumenical theory," *Appl. Opt.* **15(1)**: 84–88, Jan. 1976.

Bartell, F. O. and W. L. Wolfe, "Cavity radiation theory," *Infrared Phys.* **16**: 13–24, 1976.

Grum, F. and R. J. Becherer, *Optical Radiation Measurements*, Academic Press, New York, 1979.

Loudon, R., *The Quantum Theory of Light*, Clarendon Press, Oxford, 1973.

Nicodemus, F., *Self Study Manual on Optical Radiation Measurements, Part I, Concepts*, Superintendent of Documents, U.S. Government Printing Office, Washington, D.C. 20402, 1976.

Palmer, J. P., private communication (1978).

Papoulis, A., *Probability, Random Variables and Stochastic Processes*, McGraw-Hill, 1965.

Wolfe, W. L. and G. J. Zissis, Eds., *Infrared Handbook*, Office of Naval Research, 1978.

PROBLEMS

1-1.   Calculate the radiant photon exitance in the spectral band of 10–12 μm for a 300 K blackbody.

1-2.   Wein's displacement law gives $\lambda_m T = 2898$ μm K for blackbody radiant exitance. Why is it different for photon flux exitance?

1-3.   Prove the radiance of the image (L_e^I) equals the radiance of the source (L_e^S) for a single optical element with 100% transmission.

1-4.   Temperature contrast is defined as the derivative of the blackbody curve with respect to temperature, $[\partial M_e(\lambda, T)/\partial T]$. Derive a "Wein's displacement law" type expression for the wavelength maximum of the temperature contrast.

1-5.   The solid angle is often approximated by area (A) over the distance squared (r^2), $\Omega_a = A/r^2$:

a.   If we define the percentage error as $100 \times (\Omega_e - \Omega_a)/\Omega_e$, at which angle is the error 10%? (Ω_e is the exact expression for solid angle.)

b.   If we define $\Omega_a = \pi \theta_{1/2}^2$, at what angle is the error 10% for this approximation? ($\theta_{1/2}$ is in radians.)

1-6.   What is the advantage of plotting the blackbody curve on log–log paper?

1-7.   What is the incident photon flux density for a detector that is responsive from 0 to 10 μm and has a 60° full field of view of a 300 K background?

1-8.   Given a Lambertian radiator of area dA_s and radiance L, what is the radiant exitance M of the source into a cone of half angle $\theta_{1/2}$? What is the radiant exitance if $\theta_{1/2} = 45°$, 90°?

1-9.   Find an analytic expression approximating Planck's law for radiant exitance, $M_{e\lambda}$, when

    a. $ch/k\lambda T \gg 1$.

    b. $ch/k\lambda T \ll 1$.

    (These are Wein's approximation and the Rayleigh–Jeans law.) Graph the relative error for these approximations with respect to the Planck law for $T = 500$ K and a wavelength range of 1–12 μm.

1-10.   Convert the expression for spectral photon exitance, $M_{p\lambda}$ (q sec^{-1} cm^{-2} μm^{-1}) into an expression per unit frequency interval, that is, q sec^{-1} cm^{-2} Hz^{-1}. Do these two expressions have a maximum at the same wavelength?

1-11.   The Planck equation can be written $K \ (cap)$

$$\bar{n}_k dk = \frac{k^2}{\pi^2}(e^{hc/\lambda kT} - 1)^{-1} \, dk$$

    where $\bar{n}_k$ = the average number of photons per unit volume in the spectral interval between k and $k + dk$,

    $k = 2\pi/\lambda$,

    λ = vacuum wavelength of light,

    h = Planck's constant,

    c = speed of light in vacuum,

    K = Boltzmann's constant,

    T = absolute temperature.

    Derive this equation. (Hint: a rectangular box with perfectly conducting walls can support waves of integral numbers of half wavelengths.)

1-12.   If you wanted to prove the existence of an infrared rainbow, what would be some of the problems or considerations for designing an appropriate system?

    a.   Discuss spectral considerations.

    b.   What would the optical system look like?

   c. Discuss the radiometry including cosine falloff and predicted SNR.

1-13. How do you prove that optical system throughput is a constant? (Throughput = $A\Omega$.)

1-14. What is the expression as a function of wavelength for $A\Omega$ for a diffraction-limited system? Assume that the detector width is equal to the distance between the first zeros in the diffracted image of a point source.

1-15. Does $L/n^2 =$ constant hold for diffraction-limited systems?

1-16. An ideal photon detector has a quantum efficiency of 1 from $\lambda = 0$ to $\lambda = \lambda_m$ and has zero response at wavelengths greater than λ_m. Give an expression for the effective radiance measured by such a detector in terms of the spectral radiance of a source $L(\lambda, T)$.

1-17. The sun (6000 K blackbody) is a circular disc that subtends a diameter of 30 arc-minutes at the earth. Calculate the wavelength at which the solar spectral irradiance at the earth's surface equals the terrestrial spectral exitance at sea level. (Assume an earth temperature of 300 K and neglect atmospheric effects.)

1-18. Calculate the radiant power incident on a detector with $A_d = 1\ mm^2$, located 3 m from a 500 K blackbody with $A_s = 1\ cm^2$. Assume 300 K between the detector and blackbody source. The detector spectral response is

   a. $0-\infty$.

   b. $3-5\ \mu$m.

   c. $8-10\ \mu$m.

1-19. Over what range in temperature and wavelength is less than a 10% error introduced by ignoring the Bose factor in calculating photon noise?

1-20. Compare the photon noise due to the signal with the photon noise due to the background for the following two cases:

   a. 10.6 μm laser signal with a 300 K background,

   b. 500 K blackbody signal with a 300 K background.

   Can you describe an experiment that will differentiate between these two signals if their integrated fluxes are equal?

CHAPTER 2

INTRODUCTION TO DETECTOR PHYSICS AND NOMENCLATURE

2-1 SOLID-STATE PHYSICS

This chapter presents a summary of the physics and properties of semiconductors that is necessary for understanding the interaction of light and matter in the detection process. The detection mechanisms and noise sources that determine the performance of the more-common detector systems in use today generally involve the interaction of quantum particles with bulk matter in the solid (as opposed to gaseous or liquid) state. The particles of primary interest in detection processes are the photon and the electron.

Direct evidence for the quantum nature of light is provided by the observation that the threshold for ionizing metals by light irradiation (photoelectric effect) does not depend on the irradiance, but only on the frequency of the light. The photoelectric effect occurs only if

$$h\nu \geq \phi \tag{2.1}$$

where h is Planck's constant and was explained in the previous chapter on blackbody radiation, ν is the light frequency in hertz, and ϕ is the work function of the metal. Equation (2.1) can be interpreted as evidence for discrete photons in electromagnetic radiation, where each photon has energy

$$E = h\nu = qV \tag{2.2}$$

where the energy is expressed as electron volts, V, and q is the charge of a single electron in coulombs (1.6×10^{-19} C).

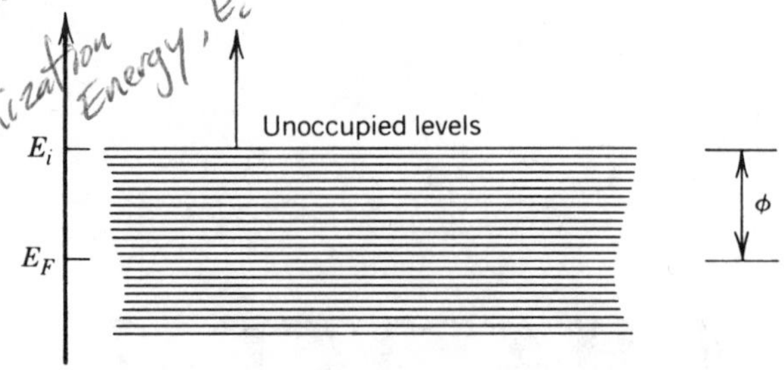

Pauli Exclusion Principle

Fermi level $P(E_F) = 50\%$

Work Function

Ionization Energy, E_i

Figure 2-1 Allowed electron energies in a metal.

The detectability of a given photon stream by a given solid-state detector is primarily determined by the relationship between the photon energies and the allowed energy states for electrons in the solid. In a solid, the allowed single-atom electron energy states are split into a large number of narrowly separated states because of the quantum mechanical restriction ("Pauli exclusion principle") that no two electrons may occupy the same state. The large number of energy levels allowed is shown in Fig. 2-1 for the case of a metal. At a temperature of absolute zero, the occupied level with the highest energy defines the Fermi level. The work function of the metal is

$$\phi = E_i - E_F \tag{2.3}$$

where E_i is the ionization energy and E_F is the energy of the Fermi level. The Fermi energy level is the energy level at which the probability of the energy state being occupied is 50% as defined by the Fermi–Dirac distribution (Sze, 1981):

$$P(E_n) = \frac{1}{\exp[(E_n - E_F)/kT] + 1} = \frac{1}{e^{\phi/kT} + 1} \tag{2.4}$$

where E_n is the energy of the nth state, k is Boltzmann's constant, and T is the temperature. The Fermi–Dirac distribution for electrons in a metal varies with temperature, as shown in Fig. 2-2.

An insulator can be differentiated from a metallic conductor by the existence of a forbidden region or band gap of several electron volts. These band gaps along the energy axis in Fig. 2-2 make up what is known as the conduction band and the valence band (see Fig. 2-2, marked as V and C on the E_n-axis). The fact that the width of the band gap (E_g in electron volts) shown in Fig. 2-3 is such that essentially no electrons are thermally excited into the conduction band characterizes the material as

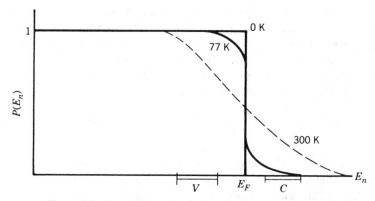

Figure 2-2 Fermi–Dirac distribution for electrons in a metal.

an insulator. The material remains an insulator as long as the thermal energy is sufficiently low; that is (Weidner and Sells, 1969)

$$\frac{kT}{q} \ll E_g \quad (eV) \tag{2.5}$$

Note that the Fermi level lies halfway between the bands for an insulator.

The class of semiconductors having a value of conductivity between that of a conductor and an insulator is known as the intrinsic semiconductor group. The name intrinsic semiconductor applies because the material is a semiconductor at the temperatures of interest without the aid of some

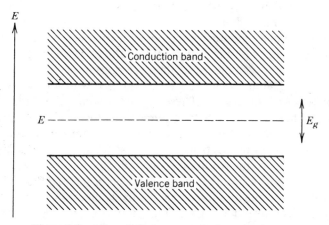

Figure 2-3 Allowed electron energies in an insulator.

process (such as adding impurities) to make the material into a semiconductor. The excitation of electrons into the conduction band leaves "holes" (unoccupied allowed states) in the valence band. The positively charged holes can carry current through the migration of valence electrons. In thermal equilibrium the number of holes is equal to the number of electrons, and the intrinsic carrier concentration can be calculated

$$N_i^2 = NP = 4\left[\frac{2\pi m k T}{\hbar^2}\right]^3 e^{-qE_R/kT} \quad \text{[handwritten: } N_i \to 0 \text{ as } T \to 0\text{]} \quad (2.6)$$

where N_i is the intrinsic carrier concentration (number of carriers per cubic centimeter), N is the concentration of negative carriers (electrons), P is the concentration of positive carriers (holes), and m is the effective carrier mass. For silicon at 300 K, N_i^2 is about 4×10^{20} cm^{-6}. The conductivity σ is the product of the number of carriers ($N = P$), the charge per carrier (q), and the mobility of the carriers (the electron mobility μ_e essentially), or

$$\sigma = Nq\mu_e \quad (2.7)$$

Therefore, an increase in the number of carriers by means of either an increase in temperature or the choice of a material with a smaller band gap will increase the conductivity. Additionally, electrons may be excited to the conduction band by the absorption of photons if the photon energy is greater than the energy gap:

$$h\nu \geq qE_g \quad (2.8)$$

Measuring the conductivity (or, equivalently, the resistance) of a semiconductor irradiated by light (photons) is one of the most common detection mechanisms in use.

Extrinsic semiconductors become semiconductive because of the presence of impurity atoms in the crystal lattice. The effect of impurities with more valence shell electrons than are present in the valence shell of the bulk crystal atoms is to add electrons that are not involved in the covalent bonds of the crystal. These impurity atoms are said to be donors of negative charge, and an N-type semiconductor results. Impurity atoms with fewer valence electrons than required for complete crystal bonding are called acceptors of electrons. The presence of acceptors implies the presence of holes, so the resultant semiconductor is said to be P type. The effect of the impurities on the energy diagram of the material is illustrated in Fig. 2-4. The impurities introduce available electron energy states into the (previously) forbidden gap. The donor impurities in an N-type semiconductor produce states just below the conduction band, in effect

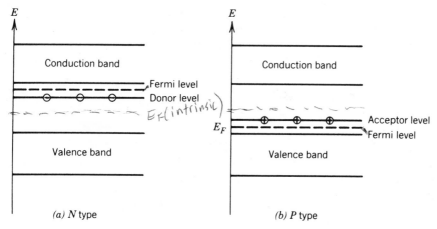

Figure 2-4   Effect of impurities on energy levels.

lowering the boundary of the conduction band and raising the Fermi level. The Fermi level is between the donor band and the conduction band because the donor and valence levels are full at absolute zero, while the conduction levels are empty. The acceptors in a *P*-type semiconductor introduce states just above the valence band and hence lower the Fermi level to the position shown in Fig. 2-4. Semiconductivity results from electron transport in *N*-type material and from hole transport in *P*-type material.

When a *P–N* junction is formed between acceptor- and donor-doped regions of a semiconductor material, the equilibrium state of the material requires that the Fermi level be at the same energy value throughout the material. This results in a transition between different energy levels for the conduction and valence bands across the junction. Figure 2-5 illustrates the resulting structure. Note that the bands have been compressed into lines to simplify the drawing. Alternatively, the figure can be viewed as presenting only the lowest-energy conduction states and the highest-energy valence states.

The difference in energy levels for the *N*- and *P*-region conduction bands results in an equilibrium potential difference ϕ_0:

$$\phi_0 \approx \frac{kT}{q} \ln\left[\frac{N_p P_n}{N_i^2}\right] \tag{2.9}$$

where N_p is the concentration of electrons in the *P* region, P_n is the concentration of holes in the *N* region, and N_i is the intrinsic carrier

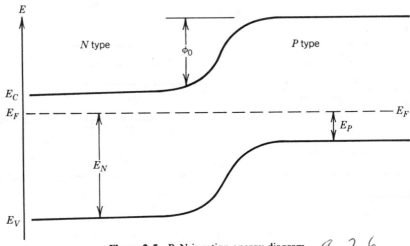

Figure 2-5   P–N junction energy diagram.

concentration. N_p results from the migration of electrons from the N-region to P-region acceptors. P_n is caused by the migration of holes from the P-region to the N-region donors. N_i is due to the thermal excitation of hole–electron pairs [Eq. (2.9)]. The presence of a potential barrier causes the P–N junction device to be a nonlinear circuit element. If an external bias voltage is applied so as to overcome the potential barrier (which typically requires on the order of 0.7 V for silicon), a very high forward current is produced. Reversing the bias voltage increases the potential barrier, and very little current flows. The characteristics of a P–N junction are utilized in photovoltaic detectors solar cells, and heterodyne detection.

The mobility of electrons (or of holes) is a parameter that can be used to quantify the useful properties of semiconductor devices. The mobilities relate the charge carrier average velocities to the material conductivity, incorporating the electron and hole effective masses.

In semiconductors, conduction-band electrons and valence-band holes respond in most cases like free carriers obeying Maxwell–Boltzmann statistics. (Electrons and holes must satisfy the Pauli exclusion principle and therefore actually obey Fermi–Dirac statistics.) The Fermi–Dirac distribution functions are more difficult to manipulate, and in the mobility development provide essentially the same result. The difference arises in evaluating the mean time between collisions (McKelvey, 1965). According to Maxwell–Boltzmann statistics, the average square of the thermal velocity of a charge carrier is given by

electron (or carrier) velocity depends upon temperature
holes

$$\tfrac{1}{2}m\langle v^2 \rangle = \tfrac{3}{2}kT \quad \text{or} \quad \langle v^2 \rangle = \frac{3kT}{m} \tag{2.10}$$

(This gives an average velocity on the order of 10^7 cm/sec for semiconductors at room temperature.) This thermal velocity appears in the expression for the average drift velocity of carriers in response to an electric field applied to the semiconductor, and is also a factor in determining the mean time between collisions, $\bar{\tau}$. To arrive at the expression for drift velocity, the Boltzmann continuity equation for the distribution function describing velocities of charges is modified to account for the presence of the electric field. Then the product of this distribution function and the density-of-states function is used as a velocity probability function, and the appropriate integration over all possible velocities gives the average drift velocity. This integration results in

$$V_d = \frac{\int v P(v) d^3 v}{\int p(v) d^3 v} = -\frac{qE_0}{3kT}\langle v^2 \tau \rangle \tag{2.11}$$

If we substitute for $\langle v^2 \rangle$ and using

$$\bar{\tau} = \frac{\langle v^2 \tau \rangle}{\langle v^2 \rangle} \tag{2.12}$$

then

$$v_d = \frac{qE_0 \bar{\tau}}{m}$$ Drift velocity as a function of voltage

The mass m appearing in the above expression is that of a free electron, which, because of the interaction of the charge carrier with the periodic potential of a semiconductor crystal, may inaccurately represent the inertial properties of the charge carrier. The correct value to use in its place is the electron (or hole) effective mass. The quantum-mechanical description of the electron in a periodic potential results in an expression for the electron effective mass $m_e^* = \hbar^2/(d^2\varepsilon/dk^2)$ and for holes, $m_p^* = \hbar^2/(d^2\varepsilon/dk^2)$. The factor $d^2\varepsilon/dk^2$ is the curvature of the conduction-band energy in k space, for electrons, and of the valence-band edge energy for holes. In semiconductors, these effective masses have values ranging from near the gravitational mass value of the electron down to only a few percent of it. It is interesting to note that because the conduction- and valence-band edges may not have the same curvatures, the electron and hole effective masses may be dissimilar. Many devices are designed to take advantage of this characteristic by doping the semiconductor so that the majority carrier is the one with the lower effective mass.

Returning to the mobility, it is defined as the ratio of the average drift velocity to the applied electric field. This results in an expression for the mobility

$$\mu = \frac{v_d}{E_0} = \frac{q}{m^*}\bar{\tau} \tag{2.13}$$

To see the connection with conductivity, consider that the current in a material is given by $I = -nq\bar{v}_d$, where n is the number of carriers. This gives

$$I = nq\left(\frac{qE_0\bar{\tau}}{m^*}\right) = nq\mu E_0 = \sigma E_0 \tag{2.14}$$

Therefore, it follows that

$$\sigma = nq\mu \tag{2.15}$$

Drift in response to applied electric field is not the sole mechanism for current flow in a semiconductor. The other mechanism is charge carrier diffusion, caused by spatial gradients in the charge carrier density. The total current densities **J** for electrons and holes can be written (McKelvey, 1965)

$$\mathbf{J}_n = -D_n \mathbf{\nabla}_n - n\mu_n \mathbf{E}$$

and

$$\mathbf{J}_p = -D_p \mathbf{\nabla}_p + p\mu_p \mathbf{E} \tag{2.16}$$

where n and p are the electron and hole carrier densities, and D_n and D_p are the diffusion constants for electrons and holes, respectively. The second term is the drift contribution discussed above. The first term is due to diffusion.

The diffusion constants are given by the expressions

$$D_n = \frac{x_n \langle v_n \rangle}{3} \quad \text{and} \quad D_p = \frac{x_p \langle v_p \rangle}{3}$$

where x_n and x_p are the mean free path lengths for electrons and holes, respectively, and $\langle v_n \rangle$ and $\langle v_p \rangle$ are the associated average velocities. McKelvey (1965) suggests that the diffusion constant expressions can be derived from the Boltzmann continuity equation. If the mean free paths and the collision free times $\bar{\tau}_n$ and $\bar{\tau}_p$ are independent of velocity, then

L (diffusion length) ≡ Distance a carrier travels before recom[bination]
Distance at which recombination occurs

$x_{n,p} = (3\pi/8)\langle v_{n,p}\rangle\bar{\tau}_{n,p}$. Noting that the average thermal velocity for a Boltzmann distribution is $\langle v\rangle = (8kT/\pi m^*)^{1/2}$, we obtain

$\bar{\tau}_n \Rightarrow$ collision-free time

$$D_{n,p} = \frac{kT\bar{\tau}_{n,p}}{m^*_{n,p}} = \frac{kT\mu_{n,p}}{q} \qquad (2.17)$$

The Einstein Relation

Diffusion Constant

This last expression, connecting the diffusion constants with the mobilities, is known as the Einstein relation.

One other parameter is conceptually useful. Because carriers have finite recombination lifetimes, and because they diffuse away from the higher-density regions where they are generated, a characteristic diffusion length is defined as the distance a generated carrier will diffuse before recombining. By solving the current continuity equation under steady-state conditions with no applied field and no excess bulk carrier generation, we find that the excess carrier density dependence on x is given by

$$\delta_n = A\exp[-x/(D_n\tau_n)^{1/2}] \quad \text{for electrons} \qquad (2.18)$$

$\tau_n =$ excess carrier lifetime

or

$$\delta_p = A\exp[-x/(D_p\tau_p)^{1/2}] \quad \text{for holes} \qquad (2.19)$$

D_n and D_p are the diffusion constants discussed earlier. τ_n and τ_p are not the collision free times, but are the excess carrier lifetimes for electrons and holes. From this expression it is clear that defining the diffusion lengths

$$L_n = \sqrt{D_n\tau_n} \quad \text{and} \quad L_p = \sqrt{D_p\tau_p} \qquad (2.20)$$

is convenient. By dimensional arguments the diffusion length can also be written

$$L_{n,p} = \frac{D_{n,p}}{\langle V\rangle_{n,p}} \leftarrow \text{mean thermal velocity} \qquad (2.21)$$

where $\langle V\rangle$ is the mean thermal velocity. By combining several of the above expressions, the mobility and the diffusion length can be related:

$$\mu = \frac{qL\langle V\rangle}{kT} = \frac{3qL}{m\langle V\rangle} \qquad (2.22)$$

Mobilities and diffusion lengths are important in semiconductive detectors. Higher mobilities indicate lower internal resistive losses and higher current responsivities. In photoconductive devices, high mobility allows the carriers to be swept from the semiconductor by the external field before recombination occurs. Longer diffusion lengths are important in photovoltaic devices, where photogenerated minority carriers must diffuse to the edge of the $P-N$ junction depletion region in order to produce an external voltage (or an external current in short-circuit or back-biased operation) for the device. Diffusion lengths in silicon are about 1 cm whereas in InSb they are about 10^{-2} cm.

2-2   DETECTION MECHANISMS

A brief introduction to some common detection mechanisms is presented. These detection mechanisms are described in more detail in later chapters.

The photovoltaic effect is the generation of a potential across a $P-N$ junction when radiation is incident on it. When photon flux irradiates the junction, electron–hole pairs are formed if the photon energy exceeds the forbidden gap energy, and the field (of magnitude ϕ_0 in Fig. 2-5) sweeps the electrons from the P region to the N region, and holes from the N region to the P region. This process makes the P region positive and the N region negative, and will produce current flow in an external circuit.

The voltage–current characteristic of a photovoltaic detector is shown in Fig. 2-6. If there is no radiation incident on the detector, the $V-I$ curve

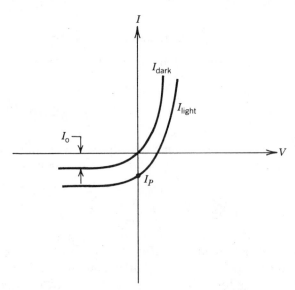

Figure 2-6   Voltage–current characteristics of a photovoltaic detector.

passes through the origin (Kittel, 1969)

$$I_{dark} = I_0(e^{qV/\beta kT} - 1) \tag{2.23}$$

The curve is shifted downward by an amount defined as the photocurrent in the presence of radiation:

$$I_{light} = I_{dark} + I_{photo} = I_p \tag{2.24}$$

for the short-circuit case ($V = 0$) illustrated in Fig. 2-6. Photovoltaic detectors are normally operated into a (virtual) short circuit because the photocurrent is then a linear function of the magnitude of the irradiance, as will be shown in Chapter 3. This linear behavior implies that the detector impedance is independent of photon flux. This can be seen from

$$R_d = \frac{\partial V}{\partial I} = \frac{\beta kT}{qI_0} e^{-qV/\beta kT} = \frac{\beta kT}{qI_0} \tag{2.25}$$

for $V = 0$. The short-circuit impedance of the detector is constant if the temperature is maintained at a constant value.

A photoconductive detector operates on the basis that the conductivity σ [Eq. (2.7)] of a semiconductor can be changed by an incident photon stream. If the photons have sufficient energy

$$h\nu \geq E_g \tag{2.26}$$

then additional electrons can be excited into the conduction band. The change in conductivity due to the increase in carriers is

$$\Delta\sigma = \Delta Nq\mu \tag{2.27}$$

where ΔN is the change in the number of conduction-band electrons, q is the change per electron, and μ is the electron mobility. The exact relation of ΔN to the irradiance and the photoconductive gain is presented in Chapter 4. The change in conductivity is used to produce an electrical signal that is dependent on the irradiance by externally applying a constant bias voltage across the photoconductor. The current in the circuit is then a function of the irradiance.

Photoemissive detectors employ the photoelectric effect. If a thin photocathode and an anode are placed in a vacuum container with an external applied voltage present (Fig. 2-7), then a current proportional to the number of incident photons will flow with sufficient energy to produce free electrons:

$$h\nu > \psi \tag{2.28}$$

where ψ is the electron affinity of the photocathode. If successive dynodes

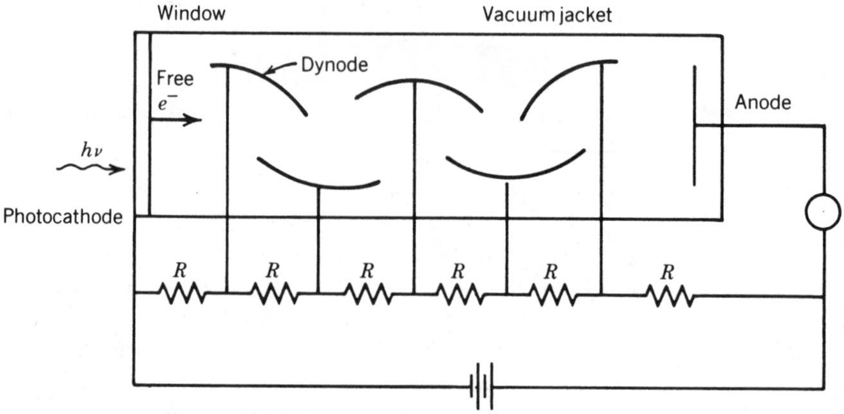

Figure 2-7   Photomultiplier tube schematic.

are introduced as shown in the figure, then a photomultiplier tube with gain is produced. A single electron accelerated through an electric potential difference V has kinetic energy qV. If

$$qV \geq \delta\phi \qquad (2.29)$$

where ϕ is the work function of the dynodes, then δ electrons can be emitted from the dynode for one incident electron. For a chain of P dynodes, an overall electron current gain on the order of

$$G = \delta^P \qquad (2.30)$$

is possible. A more precise treatment is presented in Chapter 5. Note that the gain of any detector refers only to an electron current gain. Light gain or amplification is present only in Laser (light amplification by stimulated emission of radiation) and maser (microwave amplification by stimulated emission of radiation) devices. These devices are not detectors, but amplifiers. They are usually employed as signal sources, although masers have been used in microwave receiver systems to amplify the signal prior to detection.

The bolometer is a detector whose resistance changes as a function of irradiance, but it behaves differently from a photoconductor. Figure 2-8 illustrates an important spectral response difference between the two detector types. It was stated earlier that for a photoconductor the photons must exceed a minimum energy in order to be detected, and, therefore, must be shorter than a cutoff wavelength (λ_p):

$$\lambda_p = \frac{hc}{E_g} \qquad (2.31)$$

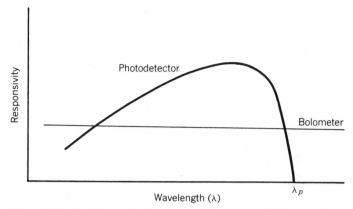

Figure 2-8   Relative spectral response of bolometer and photoconductor.

The increasing response with wavelength for the photodetector is discussed in Chapters 3 and 4. Figure 2-8 shows that while a constant bias and irradiance level produce a varying voltage output with wavelength of the radiation incident on the photodetector, the voltage output from the bolometer is constant for constant irradiance. The bolometer may be considered an energy detector or thermal detector.

The fact that the bolometer responds differently to incident radiation than a photoconductor is due to a different mode of physical operation. The bolometer may be treated as a thermistor; that is, a resistor with a large temperature coefficient (ohms per degree of temperature). The voltage drop is

$$V_d = i_{\text{bias}} \gamma \, \Delta T \tag{2.32}$$

where ΔT is the temperature change, and the temperature coefficient γ is

$$\gamma = \frac{\partial R}{\partial T} \tag{2.33}$$

If the bolometer absorbs incident radiation uniformly without wavelength dependence, then the temperature change depends only on the incident energy, and the uniform response of Fig. 2-8 will be achieved. Chapter 7 gives a more complete treatment of bolometers. Bolometers of exceptional sensitivity can be constructed by cooling the detector to low temperatures on the order of 4 K or less so that a given temperature change (say 0.01 K) represents a relatively large percentage change. The fact that the high performance is available at long wavelengths makes the bolometer a relatively common far infrared and millimeter wave (~100 μm to 1.0 mm) detector.

The thermocouple can be used as a detector of irradiant energy because of the fact that a voltage is produced when a temperature gradient exists along two dissimilar conductors that are connected at a junction. The open-circuit voltage generated is usually on the order of tens of microvolts per degree centigrade of temperature difference for a single junction. Many junctions may be connected in series to increase the voltage produced. A series connection of thermocouples is called a thermopile. The thermopile is more massive than a single thermocouple, so it must be irradiated for a longer time to yield a given temperature rise. This slow response (and hence low-frequency bandwidth) is characteristic of most thermal detectors when compared to photon detectors. The physics and detection performance of the thermocouple are described in Chapter 6.

The pyroelectric detector, described in Chapter 8, is the fastest of the common thermal detectors, with a frequency response often extending to 1 GHz. This rapid response is due to the fact that temperature changes at the molecular level are directly responsible for the detection process. The pyroelectric detector can be viewed as producing an alternating voltage signal in series with a capacitance and resistance. The detector is only responsive to energy changes that produce alternating currents of sufficiently high frequency that the capacitance presents a low impedance. The physical process that occurs in pyroelectric detection is a change in the polarization of the detector material. This change in polarization causes a change in the charge stored across the capacitor. A change in charge on the capacitor is accompanied by an output voltage change.

2-3 NOISE SOURCES

The concept of noise as an uncertainty in the *radiated* signal level was introduced in Section 1.5. This section deals with the sources of *electrical* noise fluctuations in the detection system. The noise associated with a *detector system* can be subdivided into three major classes (Fig. 2-9):

  I. Photon Noise
     A. Noise due to the signal radiation
     B. Noise due to background radiation
  II. Detector-Generated Noise
     A. Johnson Noise
     B. Shot Noise

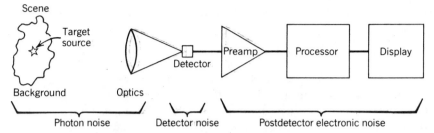

Figure 2-9   Detector system noise classification.

    C. Generation–Recombination Noise
    D. $1/f$ Noise
    E. Temperature Fluctuations
    F. Microphonics
III. Post-detector Electronic Noise
    A. Johnson Noise
    B. Shot Noise
    C. Generation–Recombination Noise
    D. $1/f$ Noise
    E. Temperature Fluctuations
    F. Popcorn Noise
    G. Microphonics

First, a general statistical consideration for noise will be discussed. Noise power can be treated as a variance σ^2 associated with the standard deviation σ corresponding to the voltage or current noise. Both treatments yield the same results. For a voltage noise source, then, the noise σ is

$$\sigma = \sqrt{\sigma^2} = [\overline{\Delta V^2}]^{1/2} = \{\overline{[V - \bar{V}]^2}\}^{1/2}$$
$$= \left\{ \frac{1}{T} \int_0^T [V(t) - \bar{V}]^2 \, dt \right\}^{1/2} \tag{2.34}$$

where $\bar{V}$ is the average signal voltage and T is the integration time or electrical time constant of the detector system. The total noise from n sources is

$$\sigma_{\text{tot}} = \left[\sum_{i=1}^{n} \sigma_i^2 \right]^{1/2} = \Delta V_{\text{rms}} = \left[\sum_{i=1}^{n} \overline{\Delta V_i^2} \right]^{1/2} \tag{2.35}$$

where ΔV_{rms} is known as the root-mean-square total voltage noise:

$$\Delta V_{rms} = \left[\frac{1}{nT} \sum_{i=1}^{n} \int_{0}^{T} [V_i(t) - \bar{V}]^2 \, dt \right]^{1/2} \tag{2.36}$$

If the noise is represented as a stationary ergodic random process, some Fourier techniques can be used to analyze it. Because the output voltage fluctuates in time owing to nonperiodic generation rates, we can look at these fluctuations in terms of the noise power spectrum or spectral power density.

An autocorrelation $[\mathscr{R}_{vv}(\tau)]$ of the noise is shown in Fig. 2-10a. The dc component is assumed to be zero, and τ is the shift in time. The Fourier transform of this autocorrelation is the power spectral density (Gaskill, 1978).

From Fourier transform relations,

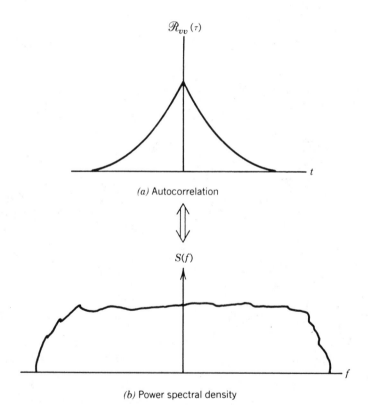

$\mathscr{R}_{vv}(\tau)$

(a) Autocorrelation

$S(f)$

(b) Power spectral density

Figure 2-10 Fourier transform relation between autocorrelation and power spectral density: (a) autocorrelation; (b) power spectral density.

$$\mathcal{R}_{vv}(0) = \int_{-\infty}^{\infty} S(f)\,df \qquad (2.37)$$

But

$$\mathcal{R}_{vv}(0) = \int_{-\infty}^{\infty} V^2(t)\,dt \qquad (2.38)$$

where the autocorrelation at $\tau = 0$, or unshifted, equals the square of the voltage fluctuations. If we average this over some time T, then we can say that the mean square average is the variance. This variance is also the integral of the power spectral density over Δf, where

$$\Delta f = \frac{1}{2T} \qquad (2.39)$$

2-3-1 Photon Noise

Photon noise was discussed in detail in Chapter 1. The important result is that the noise in an incident photon stream expected to contain $\bar{n}$ photons in a given time period Δt is

$$\sigma = \sqrt{\bar{n}} = \sqrt{\phi_P \Delta t} \qquad (2.40)$$

since the Bose correction [Eq. (1.61)] can be ignored for most optical cases.

2-3-2 Johnson Noise

Johnson or Nyquist noise is caused by the thermal motion of charged particles (thermal current fluctuations) in a resistive element. The measurable noise energy is equal to one-half the thermal energy $(\frac{1}{2}kT)$ because of the equipartition of energy between the noise source and measurement device. The rate of noise energy transfer through a circuit of time constant Δt is then the observable Johnson noise power:

$$P_{J\,\text{rms}} = \frac{kT}{2\Delta t} = kT\,\Delta f \qquad (2.41)$$

when Δf is the effective bandwidth (Hz) of a circuit that is equal to the reciprocal of twice the time constant [Eq. (2.39)].

The Johnson rms noise voltage must be the voltage that will cause the noise power of Eq. (2.41) to be transferred from the noise-producing resistor to another resistor of equal value connected in parallel (since maximum power transfer occurs when impedances are matched). The power transferred from R_1 to R_2 connected in parallel is

$$P_{12} = \frac{R_2}{(R_1 + R_2)^2} V_{J\,rms}^2 \tag{2.42}$$

Setting $R_1 = R_2 = R$ for impedance matching to maximize power transfer:

$$P_{12} = \frac{R}{(2R)^2} V_{J\,rms}^2 = P_{J\,rms} \tag{2.43}$$

Solving for the Johnson noise voltage yields

$$V_{J\,rms} = \sqrt{4kTR\,\Delta f} \tag{2.44}$$

A similar procedure can be used to verify that the Johnson noise current is

$$i_{J\,rms} = \sqrt{4kT\,\Delta f / R}. \tag{2.45}$$

2-3-3 *Shot Noise*

The shot noise generated in a photon detector is due to the discrete nature of photoelectron generation. It was stated in Chapter 1 that if the arrival of incident photons is Poisson distributed, then the generation of photoelectrons is also Poisson distributed. The average current flow from an irradiated photodetector is

$$\bar{i} = \frac{\bar{n}q}{\tau} \tag{2.46}$$

where $\bar{n}$ is the average number of photoelectrons produced in a time period τ.

The instantaneous current flow is

$$i = \frac{nq}{\tau} \tag{2.47}$$

Therefore, one can write an expression for the variance (shot noise current squared) of this current as

$$i_{sn\,rms}^2 = \overline{(i - \bar{i})^2} = \overline{\left(\frac{nq}{\tau} - \frac{\bar{n}q}{\tau}\right)^2} = \frac{q^2}{\tau^2}\overline{(n - \bar{n})^2} \tag{2.48}$$

From Poisson statistics the variance $\overline{(n - \bar{n})^2}$ of the number of electrons is just the mean $\bar{n}$. Therefore, from Eq. (2.40) and using Eqs. (2.39) and (2.46),

$$i_{sn\,rms}^2 = \frac{q^2\bar{n}}{\tau^2} = \frac{q\bar{i}}{\tau} = 2q\bar{i}\,\Delta f$$

$$i_{sn\,rms} = \sqrt{2q\bar{i}\,\Delta f} \tag{2.49}$$

which is the value of shot noise for photon detectors irradiated by a Poisson-distributed photon stream. Equation (2.49) and the term shot noise should be applied only to photon detectors that contain a potential barrier such as photovoltaic detectors. These detectors are characterized by a single degree of freedom, that is, they are diodes that conduct current in only one direction as a circuit element. Photoconductors have no preference in current flow direction and have two degrees of freedom. The photon contribution to generation–recombination (G-R) noise in a photoconductor is $\sqrt{2}$ higher in magnitude than the value given by Eq. (2.49). Since recombination does not occur in ideal diodes, it is common to separate those detectors that are said to exhibit shot noise (e.g., photovoltaic) from those that exhibit G-R noise (e.g., photoconductors). It would be possible to formulate a general parametric expression for G-R noise that applied to all photon detectors, choosing parameter values appropriate to the detector in question. The important result is that one should not include shot noise as an additional source of noise if one has already calculated the G-R noise.

2-3-4 *Generation–Recombination*

The generation–recombination noise in a photoconductor is caused by the fluctuations in current carrier generation (whether due to photons, thermal generation, or any other cause) and recombination. The magnitude of the noise is

$$i_{\text{G-R rms}} = 2\,\bar{i}\left[\frac{\tau_c\,\Delta f}{N_0[1 + (2\,\pi f\tau_c)^2]}\right]^{1/2} \tag{2.50}$$

where $\bar{i}$ is due to all sources of current carriers (not just photoelectrons), τ_c is the carrier lifetime, N_0 is the total number of free carriers, and f is the frequency at which the noise is measured. In the case of photoconductors, photon noise manifests itself as G-R noise. It will be shown in Chapter 4 that Eq. (2.50) can be written

$$i_{\text{G-R rms}} = 2qG(\eta E_p A_d \Delta f)^{1/2} \tag{2.51}$$

where G is the photoconductive gain (number of electrons generated per photogenerated electrons).

2-3-5 *1/f Noise*

Perhaps the most actively studied but least well understood noise source is called "1/f noise," because its power spectrum falls off rapidly with frequency (1/f). It is often stated that the cause is due to a lack of ohmic contact at the detector electrodes to surface-state traps or dislocations, but

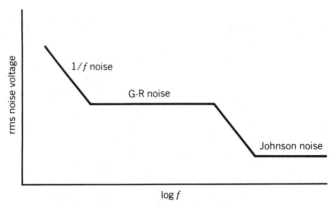

Figure 2-11 Photoconductor noise spectrum.

that is probably not a complete description (Hooge, Kleinpenning, and Vandamme, 1981). It is known that lack of ohmic contact increases the magnitude of the noise, which is different for each individual detector. It is not known if any type of physical connection can eliminate the noise source, however. The empirical expression for the noise current is

$$i_{\text{rms}} = a\sqrt{\frac{i_b^{\alpha}\Delta f}{f^{\beta}}} \tag{2.52}$$

where a is a proportionality constant, i_b is the current through the detector, α is usually found to be about 2, and β is usually about 1. The current is of course not infinite at direct current ($f = 0$). The formula merely indicates the generally observed behavior of a noise mechanism (or set of mechanisms) that is not understood. This $1/f$ noise is always present in photoconductors and bolometers that require current bias to operate. Photovoltaic detectors operated at zero bias current do not exhibit this noise. A typical photoconductor noise spectrum is shown in Fig. 2-11. The best performance (highest signal-to-noise ratio) is generally obtained by operating the photoconductor in the frequency range of generation–recombination noise dominance. Operation in this region is achieved by modulating the signal (either at the source or at the sensor) so that it falls in the desired band after detection, and by using an electrical filter in the postdetector electronics that rejects noise outside the unused frequency spectrum. The bandwidth required depends on the application.

2-3-6 *Temperature Noise*

In thermal detection systems, any fluctuations in the temperature of the detector element that are not due to a change in the signal will produce unwanted noise in the electrical output. This noise source is called temperature noise to distinguish it from thermal (Johnson) noise and is sometimes called phonon noise. The temperature fluctuations due to radiative and conductive heat exchange result in a fluctuation noise:

$$T_{rms} = \left\{ \frac{4Kk\bar{T}^2 \Delta f}{K^2 + [2\pi fH]^2} \right\}^{1/2} \tag{2.53}$$

where K is the thermal conductance and H is the heat capacity. The resulting noise voltage on current across the detector terminals is determined by the type of thermal detector in use. The time constant for this noise is determined by

$$\tau_T = \frac{H}{K} \tag{2.54}$$

2-3-7   *Microphonics*

Microphonic noise is caused by the mechanical displacement of wiring and components when the system is subjected to vibration or shock. Changes in capacitance due to the change in wire spacing relative to the grounded metal case are primarily responsible for this noise. It is system dependent, and no attempt to derive a general quantitative noise expression is possible. Operationally, a system should be constructed in a mechanically stable way. If it is observed that mechanical shock modulates the system output, the mechanical arrangement of electrical components should be changed.

2-3-8   *Postdetector Electronic Noise*

Postdetector electronics exhibit each of the noise sources listed in Section 2-3. The overall noise of the electronics will usually be treated as a single measured quantity.

2-4   FIGURES OF MERIT

Figures of merit are used to compare the measured performance of one detector against the measured performance of other detectors of the same class, the performance required to perform a given task, or the calculated performance expected of an ideal detector that performs at a level limited by some fundamental physical principle. Care in the use of detector figures of merit is essential because many parameters of detector performance do not fully summarize the relevant factors in detector choice. The developer of detector systems should therefore understand the definitions and limitations of the commonly used figures of merit.

A basic figure of merit that applies to all detectors with electrical output is responsivity. Responsivity is the ratio of the output (usually in amperes or volts) to the radiant input (in watts). For example, the spectral voltage responsivity of a detector at a given wavelength λ is the measured voltage output, V_s, divided by the spectral radiant power incident on the detector, $\phi_e(\lambda)$:

$$\mathcal{R}_V(\lambda, f) = \frac{V_s}{\phi_e(\lambda)} \tag{2.55}$$

The blackbody responsivity $R(T, f)$, alternatively, is the detector output divided by the incident radiant power from a blackbody source of temperature T modulated at a frequency f that produces the observed output:

$$\mathcal{R}_V(T, f) = \frac{V_s}{\displaystyle\int_0^\infty \phi_e(\lambda)\,d\lambda} = \frac{V_s}{A_s \sigma_e T^4 A_d / \pi R^2} \tag{2.56}$$

for the case of a detector of area A_d, irradiated by a blackbody (temperature $= T$) of area A_s at a distance R. It is implicit in Eq. (2.56) that the source and detector are both normal to the optical axis. Equation (2.56) also holds if the detector is preceded by a lossless optical system ($\tau = 1$) that images all of the source area onto the detector area. Note that the blackbody responsivity is a measure of the detector response to the incident radiation integrated over all wavelengths, even though the detector is sensitive to only a finite wavelength interval.

The noise equivalent power (NEP) of a detector is the required power incident on the detector to produce a signal output equal to the rms noise output. Stated another way, the NEP is the signal level that produces a signal-to-noise ratio of 1. The current signal output is

$$i_s = \mathcal{R}_i \phi_e \tag{2.57}$$

so the signal-to-noise ratio is

$$S/N = \frac{\mathcal{R}_i \phi_e}{i_{rms}} \tag{2.58}$$

The NEP is the incident radiant power, ϕ_e, for a signal-to-noise ratio of 1:

$$1 = \frac{\mathcal{R}_i \, NEP}{i_{rms}} \tag{2.59}$$

Solving for NEP gives

$$\text{NEP} = \frac{i_{\text{rms}}}{\mathcal{R}_i} \qquad (2.60)$$

where i_{rms} is the root-mean-square noise current in amperes and $\mathcal{R}_i$ is the current responsivity in amperes per watt. Either the spectral responsivity $\mathcal{R}_i(\lambda, f)$ or the blackbody responsivity $\mathcal{R}_i(T, f)$ may be inserted into Eq. (2.60) to define two different noise equivalent powers. The spectral NEP, $\text{NEP}(\lambda, f)$, is the monochromatic radiant flux $\phi_e(\lambda)$ required to produce an rms signal-to-noise ratio of 1 at a frequency f. The blackbody NEP, $\text{NEP}(T, f)$, is the blackbody radiant flux required to produce an rms signal-to-noise ratio of 1.

The noise equivalent power is a useful parameter for comparing similar detectors that operate under identical conditions, but it should not be used as a summary measure of detector performance for comparing dissimilar detectors. For example, the noise expressions given in Section 2-3 indicate that the larger the bandwidth Δf, the larger the noise that is present. This would increase the NEP. Equation (2.56) implies that increasing the detector area A_d will decrease the responsivity if all other factors are held constant. From Eq. (2.60) it may be concluded that a decrease in responsivity will increase the NEP. It is seen, then, that both A_d and Δf influence the NEP in the same direction. However, neither the detector area nor the bandwidth was specified in the definition of NEP. A comparison of noise equivalent powers measured under different conditions can therefore be misleading.

The detectivity D of a detector is simply the reciprocal of the noise equivalent power:

$$D = \frac{1}{\text{NEP}} \qquad (2.61)$$

The only difference between NEP and detectivity is that a more-sensitive detector (one that can detect a smaller signal) has a larger detectivity than a less-sensitive detector, whereas the NEP is smaller for the more-sensitive detector. Detectivity follows the saying that "bigger is better." A more-useful figure of merit is the normalized detectivity D^* (pronounced "dee-star"), which normalizes the detector area and bandwidth:

$$D^* = D\sqrt{A_d \Delta f} = \frac{\sqrt{A_d \Delta f}}{\text{NEP}} \qquad (2.62)$$

where either spectral or blackbody NEP may be used to define spectral or blackbody detectivity, that is, $D^*(\lambda, f)$ or $D^*(T, f)$. The advantage of D^*

as a figure of merit is that it is normalized to an active detector area of $1\,\text{cm}^2$ and noise bandwidth of 1 Hz. It will become evident in later chapters that for most detectors, the signal-to-noise ratio is inversely proportional to $\sqrt{A_d \Delta f}$. Therefore, D^* may be used to compare directly the merit of detectors of different size whose performance was measured using different bandwidths. Alternative equivalent expressions to Eq. (2.62) include

$$D^* = \frac{\sqrt{A_d \Delta f}}{V_{\text{rms}}} \mathscr{R}_V = \frac{\sqrt{A_d \Delta f}}{i_{\text{rms}}} \mathscr{R}_i = \frac{\sqrt{A_d \Delta f}}{\phi_e} (\text{S/N}) \qquad (2.63)$$

where ϕ_e is radiant power incident on the detector. One way of interpreting D^* is that it is equal to the signal-to-noise ratio at the output of the detector when 1 W of radiant power is incident on a detector area of $1\,\text{cm}^2$ with a bandwidth of 1 Hz (D^* is expressed in $\text{cm}\,\text{Hz}^{1/2}\,\text{W}^{-1}$). This interpretation is meant to be used as a mental concept only, because most detectors reach their limiting output at well below 1 W of incident power, and can never achieve a signal-to-noise ratio approaching the value of D^*.

The relation between the spectral D^* and blackbody D^* is

$$D^*(T, f) = \frac{\int_0^\infty D^*(\lambda, f)\phi_e(T, \lambda)\,d\lambda}{\int_0^\infty \phi_e(T, \lambda)\,d\lambda} = \frac{\int_0^\infty D^*(\lambda, f)E_e(T, \lambda)\,d\lambda}{\int_0^\infty E_e(T, \lambda)\,d\lambda}$$

$$(2.64)$$

The far right-hand expression of Eq. (2.64) can be calculated without knowing the detector area as long as $D^*(\lambda, f)$ is known for all λ. Note that the finite wavelength interval for the spectral response of any real detector ensures that

$$D^*(\lambda_p, f) > D^*(T, f) \qquad (2.65)$$

where λ_p is the wavelength of the peak or highest value of spectral D^*.

The most frequently encountered special case of the general definitions of figures of merit is the form that describes photon-limited performance. In the visible ($0.4 - 0.7\ \mu\text{m}$) and near infrared (about $0.7-1.1\ \mu\text{m}$ in this case), photomultiplier tubes and microchannels (see Chapter 5) often attain signal photon-limited performance. In the thermal infrared ($\lambda > 3\ \mu\text{m}$), photoconductive, photovoltaic, and bolometric detectors can achieve background photon-limited performance, or in cases where the signal source fills the detector apparent field of view, signal photon-limited performance. An infrared photon detector that achieves background-limited performance is termed BLIP (background-limited infrared photodetector). The performance in all of these cases is limited by the uncertainty in arrival rate of incident photons (photon noise). The term BLIP is sometimes applied to any photon-limited infrared detector even when it is signal limited. This practice may result from the fact that the signal and background are often both blackbody sources not far apart in temperature, so physically the processes are the same. The unique characteristic of BLIP performance equations is that the photon source is the dominant source of noise. Only when most of the field of view contains background and not signal photons is the term BLIP truly appropriate.

The electrical output of a photovoltaic detector that achieves photon-limited performance contains shot noise as the dominant noise source:

$$i_{rms} = (2q\bar{i}\,\Delta f)^{1/2} \tag{2.66}$$

In this case the average output current is

$$\bar{i} = \eta[\phi_P^S + \phi_P^B]q \tag{2.67}$$

where η is the quantum efficiency (probability that a photoelectron is produced when a photon is incident on the detector), ϕ_P^S is the number of incident signal photons per second, ϕ_P^B is the number of incident background photons per second, and q is the charge carried by one photoelectron. For BLIP operation, $\phi_P^B \gg \phi_P^S$; therefore,

$$S/N = \frac{i_{s\,rms}}{i_{rms}} = \frac{\eta\phi_P^S q}{[2q^2\eta\phi_P^B\Delta f]^{1/2}} \tag{2.68}$$

By setting the rms signal-to-noise ratio equal to unity, one can obtain the noise equivalent photon flux:

$$\phi_{P\,rms}^S = \sqrt{\frac{2\phi_P^B\Delta f}{\eta}} \tag{2.69}$$

for the BLIP case. The spectral noise equivalent power is therefore

$$NEP_{BLIP}(\lambda, f) = \frac{hc}{\lambda}\sqrt{\frac{2\phi_P^B\Delta f}{\eta}} \tag{2.70}$$

From the definition of $D^*(\lambda, f)$:

$$D_{BLIP}^*(\lambda, f) = \frac{\sqrt{A_d\Delta f}}{NEP_{BLIP}(\lambda, f)} = \frac{\lambda}{hc}\sqrt{\frac{\eta A_d}{2\phi_P^B}} \tag{2.71}$$

We note that $E_P^B = \phi_P^B/A_d$, so

$$D^*_{\text{BLIP}}(\lambda, f) = \frac{\lambda}{hc}\sqrt{\frac{\eta}{2E^B_P}} \qquad (2.72)$$

Any shot noise-limited photon detector will have a D^* given by Eq. (2.72) if E^B_P is replaced by the appropriate notation for the dominant source of photon irradiance (or in general $E^B_P + E^S_P$). The significance of Eq. (2.72) for photon-limited performance is that the detector performance is limited to this value for D^* by noise in the incident photons. It has been assumed that all noise sources inherent in the detector are negligible, yet there is still a fundamental limit in physical principle to the performance that the detector may achieve. If the background is reduced to insignificance (possible in some cases), then the signal photon stream still presents a fundamental limit to detector performance.

The incident photon flux density due to the background is related to the photon flux radiance of the background in the same way as the irradiance was found in Chapter 1 [Eq. (1.42)] to be related to the radiance (L^B_P):

$$E^B_P = \pi L^B_P \sin^2 \theta_{1/2} \qquad (2.73)$$

where $\theta_{1/2}$ is the half-angle field of view of the detector. For a Lambertian source we have [similar to Eq. (1.40)]:

$$M^B_P = \pi L^B_P \qquad (2.74)$$

Therefore, for the case of an infrared photon detector whose field of view is filled by a background at temperature T_B,

$$E^B_P = \sin^2 \theta_{1/2} \int_0^{\lambda_{\max}} M_P(\lambda, T_B)d\lambda \qquad (2.75)$$

where one must remember that the photon-detection mechanisms limit the wavelength spectral range of the detector wavelength response from zero to maximum wavelength response, $\lambda_{\max}$:

$$\lambda_{\max} = \frac{hc}{E_g} \qquad (2.76)$$

where E_g is the energy to excite a photon across the energy gap in a photovoltaic or photoconductive detector, or the work function in a photoemissive detector. The fundamental limit for BLIP operation can now be calculated from

$$D^*_{\text{BLIP}}(\lambda, f) = \frac{\lambda}{hc}\left(\frac{\eta}{2\sin^2 \theta_{1/2} \int_0^{\lambda_{\max}} M_P(\lambda, T_B)d\lambda}\right)^{1/2} \qquad (2.77)$$

Appendix C contains tabulated values of integrated photon flux exitance for a variety of temperatures and cutoff wavelengths. It can be seen that the field of view $(2\theta_{1/2})$ must be specified if $D^*_{\text{BLIP}}(\lambda, f)$ is to have an unambiguous meaning since

$$D^*_{\text{BLIP}}(\lambda, f, \theta_{1/2}) = \frac{1}{\sin \theta_{1/2}} D^*_{\text{BLIP}}(\lambda, f, \pi/2) \tag{2.78}$$

This leads to the definition of a special figure of merit $D^{**}(\lambda, f)$, which applies *only* to the BLIP case:

$$D^{**}(\lambda, f) = \sin \theta_{1/2} D^*_{\text{BLIP}}(\lambda, f) \tag{2.79}$$

Using D^{**} ("dee double-star") it is possible to directly compare detectors achieving BLIP conditions without reference to the field of view.

A photoconductive detector exhibits generation–recombination noise rather than shot noise in the photon-limited case. This leads to an increase in the rms noise level and a decrease in D^*:

$$D^*_{\text{BLIP}}(\lambda, f) = \frac{\lambda}{2hc}\left(\frac{\eta}{E^B_P}\right)^{1/2} = \frac{\lambda}{2hc \sin \theta_{1/2}}\left(\frac{\eta}{\int_0^{\lambda_{\max}} M_P(\lambda, T_B)d\lambda}\right)^{1/2} \tag{2.80}$$

where Eq. (2.80) applies only to photoconductors under BLIP operation. Equation (2.80) for photoconductors and Eqs. (2.72) and (2.77) for photovoltaic detectors assume that the detector is a purely passive receiving element that does not contribute to the background radiation. This is generally true since detectors that achieve BLIP operation must be cooled well below room temperature. For cold backgrounds, such as space applications, the radical in Eq. (2.80) can become very large. As the photon flux decreases, the detector no longer is background limited and some other noise such as Johnson or preamplifier noise becomes dominant.

The discussion of BLIP spectral detectivity is now broadened to find an expression for BLIP backbody detectivity. $D^*_{\text{BLIP}}(T, f)$ as a function of $D^*_{\text{BLIP}}(\lambda_P, f)$, where λ_P is the wavelength of peak detectivity, which is also the cutoff wavelength for an ideal photon detector.

The average number of background photons of wavelength λ per sample time interval τ incident on the detector is

$$\bar{n} = \tau A_d E_P(T_P, \lambda) \tag{2.81}$$

The uncertainty in $\bar{n}$ (photon fluctuation) is

$$n_{\text{rms}} = \left[\frac{\bar{n}e^x}{e^x - 1} \right]^{1/2} \tag{2.82}$$

for a Bose–Einstein radiator [see Eq. (1.61)], where $x = hc/\lambda k T_B$. Then rms fluctuation in photoelectron generation number per second due to background photons within the total spectral response interval of the detector is

$$n_{e_{\text{rms}}} = \left[\int_0^{\lambda_P} \eta A_d E_P(T_B, \lambda) \left(\frac{e^x}{e^x - 1} \right) d\lambda \right]^{1/2} \tag{2.83}$$

Assuming that the quantum efficiency is constant with wavelength, and that the noise bandwidth due to the carrier transit time τ is

$$\Delta f = \frac{1}{2\tau} \tag{2.84}$$

we have

$$n_{e_{\text{rms}}} = \left[2\,\Delta f\, \eta A_d \int_0^{\lambda_P} M_P(T_B, \lambda) \sin^2 \theta_{1/2} \left(\frac{e^x}{e^x - 1} \right) d\lambda \right]^{1/2} \tag{2.85}$$

The number of signal photoelectrons generated due to a blackbody signal source of temperature T_S is

$$\bar{n}_S = \tau \eta A_d \int_0^{\lambda} M_P(\lambda, T_s) \frac{A_S}{R^2} d\lambda \tag{2.86}$$

The total blackbody irradiance due to the signal source is

$$E_e^{\text{BB}} = \varepsilon \sigma_e T_S^4 \frac{A_S}{R^2} \tag{2.87}$$

E_P^{BB} is the irradiance on the detector due to an exitance (M_P^{BB}) of a blackbody at temperature T_S. Multiplying the right-hand side of Eq. (2.86) by one will not change its value:

$$\bar{n}_S = \tau \eta A_d \int_0^{\lambda_P} M_P(\lambda, T_S) d\lambda \times \frac{E_e^{\text{BB}}}{\varepsilon \sigma_e T^4} \tag{2.88}$$

We now restrict the discussion to the case S/N = 1 in order to find the NEP. Setting the signal of Eq. (2.88) equal to the noise of Eq. (2.85) gives

$$\frac{E_e^{\text{BB}}}{\varepsilon} A_d \frac{\tau \eta}{\sigma_e T_S^4} \int_0^{\lambda_P} M_P(\lambda, T_S) d\lambda$$
$$= \left[\frac{1}{2\,\Delta f} \eta A_d \int_0^{\lambda_P} M_P(T_B, \lambda) \sin^2 \theta_{1/2} \left(\frac{e^x}{e^x - 1} \right) d\lambda \right]^{1/2} \tag{2.89}$$

Note that $E_e^{BB} A_d$ is the signal power on the detector. Since we have assumed S/N = 1, then

$$\text{NEP}_{\text{BLIP}}(T_S, f) = \frac{E_e^{BB}}{\varepsilon} A_d \qquad (2.90)$$

Therefore, from Eq. (2.89), the BLIP blackbody NEP is

$$\text{NEP}_{\text{BLIP}}(T_S, f) = \frac{\sigma_e T_S^4 \left[2\,\Delta f \eta A_d \int_0^{\lambda_P} M_P(T_B, \lambda) \sin^2 \theta_{1/2} \left(\frac{e^x}{e^x - 1} \right) d\lambda \right]^{1/2}}{\eta \int_0^{\lambda_P} M_P(T_S, \lambda) d\lambda} \qquad (2.91)$$

This is the BLIP expression because the noise source is the background photon fluctuations. One now can find the BLIP backbody D^* from the definition

$$D_{\text{BLIP}}^*(T_S, f) = \frac{\sqrt{A_d \Delta f}}{\text{NEP}_{\text{BLIP}}(T_S, f)} \qquad (2.92)$$

The complete expression for the common case in which $hc/\lambda \gg kT_B$ so that $e^x/(e^x - 1) \approx 1$ is

$$D_{\text{BLIP}}^*(T_S, f) = \left(\frac{\eta}{2 \int_0^{\lambda_P} M_P(T_B, \lambda) \sin^2 \theta_{1/2}\, d\lambda} \right)^{1/2} \frac{\int_0^{\lambda_P} M_P(T_S, \lambda) d\lambda}{\sigma_e T_S^4} \qquad (2.93)$$

Examination of Eq. (2.77) should convince the reader that the first term on the right-hand side of Eq. (2.93) is the peak spectral D^* for a 2π steradian field of view (if $\theta_{1/2} = \pi/2$) if we multiply and divide it by λ_P/hc:

$$D_{\text{BLIP}}^*(T_S, f) = \frac{\lambda_P}{hc} \left(\frac{\eta}{2 E_P^B} \right)^{1/2} \frac{hc}{\lambda_P} \frac{\int_0^{\lambda_P} M_P(T_S, \lambda) d\lambda}{\sigma_e T_S^4} \qquad (2.94)$$

where

$$E_P^B = \int_0^{\lambda_P} M_P(\lambda, T_B) \sin^2 \theta_{1/2}\, d\lambda$$

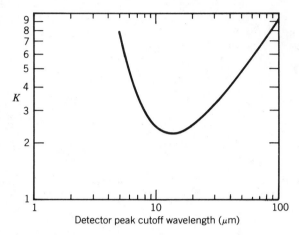

Figure 2-12   Ratio of peak spectral D^* to blackbody D^* versus detector cutoff wavelength.

or more explicitly

$$D^*_{\mathrm{BLIP}}(T_S, f) = D^*(\lambda_P, f)\left[\frac{(hc/\lambda_P)\displaystyle\int_0^{\lambda_P} M_P(T_S, \lambda)\,d\lambda}{\sigma_e T_S^4}\right] \qquad (2.95)$$

the right-hand side being the peak spectral D^* multiplied by the fraction of total energy emitted by the signal blackbody source contained in the spectral response interval of the detector. This fraction is always less than one for $\lambda_P < \infty$, which agrees with Eq. (2.65):

$$D^*_{\mathrm{BLIP}}(\lambda_P, f) > D^*_{\mathrm{BLIP}}(T_S, f) \qquad (2.96)$$

The ratio of the BLIP peak spectral D^* to the BLIP blackbody D^* is

$$K(T, \lambda) = \frac{D^*_{\mathrm{BLIP}}(\lambda_P, f)}{D^*_{\mathrm{BLIP}}(T_S, f)} = \frac{\sigma_e T_S^4}{\dfrac{hc}{\lambda_P}\displaystyle\int_0^{\lambda_P} M_P(T_S, \lambda)\,d\lambda} \qquad (2.97)$$

Figure 2-12 is a plot of $K(\lambda)$ for $T_S = 500$ K and a 2π steradian field of view. The quantity $K(T, \lambda)$ is useful because infrared detector testing yields blackbody D^* values. Peak spectral D^* is then calculated using $K(T, \lambda)$.

2-5   DETECTOR PERFORMANCE MEASUREMENT

The methodology for measuring the figures of merit defined in the previous section is presented next. The intent of the figures of merit is to

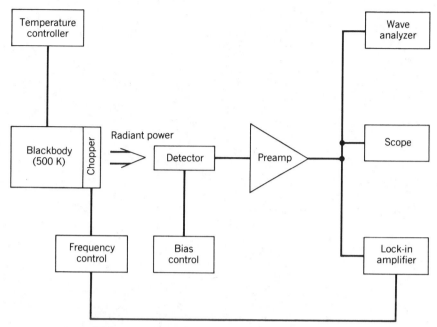

Figure 2-13   Blackbody detector testing configuration.

permit calculation of the signal-to-noise ratio which will be achieved by a working detector system. It should not be surprising, then, that the detector test methodology consists, in essence, of measuring the signal-to-noise ratio achieved under documented conditions.

A typical test configuration for measuring detector performance with a blackbody source is illustrated in Fig. 2-13. The blackbody source is usually 500 K for a thermal infrared detector test, and 2870 K is used for visible and near infrared detectors. The variable-speed chopper modulates the signal at a frequency f by rotating a notched wheel in front of the source. The notches alternately cover and uncover the source, producing a nearly square-wave signal if the source aperture is small compared to the notch width. The detector is located at a known distance from the source so that the signal on the detector can be calculated. The detector bias for the optimum signal-to-noise ratio must be found experimentally for each individual detector. An amplifier of known gain, noise level, and frequency response is used to provide the signal level required by the wave analyzer. The wave analyzer is an rms voltmeter that has a tunable bandpass filter set to a center frequency of f with bandwidth Δf. A lock-in

Room Temperature (K)	Bias (V)	Signal (mV)	Noise (μV)	Δf (Hz)	f (Hz)	A_d (mm²)	Det. FOV	Blackbody Source		
								Range (m)	Temperature (K)	Diameter of Aperture (cm)
300	90	30	30	10	1000	1	20°	2.5	500	1

Figure 2-14 Sample data sheet for an infrared photoconductor.

voltmeter is often used in place of the wave analyzer and provides synchronous detection of the electrical signal at frequency f. An example of the data required to specify the performance of an infrared photodetector is given in Fig. 2-14.

The rms signal incident on the detector in Fig. 2-13 is

$$\phi_e^S = \frac{L_e^S A_{BB} A_d}{R^2} \tau F_F \tag{2.98}$$

where L_e^S is the signal radiance, A_{BB} is the area of the blackbody source aperture, A_d is the detector area, R is the separation between source and detector, τ is the transmission of the intervening atmosphere and optical windows, and F_F is the form factor from peak-to-peak signal to rms signal. The rms radiant signal must be calculated because the resulting electrical signal from the detector is measured by an rms voltmeter. The signal radiance is

$$L_e^S = \frac{\varepsilon_{BB} \sigma_e T_{BB}^4}{\pi} - \frac{\varepsilon_{rm} \sigma_e T_{rm}^4}{\pi} \tag{2.99}$$

where ε_{BB} is the emissivity of the source, T_{BB} is the temperature of the source, ε_{rm} is the weighted average emissivity of the chopper wheel and room environment in the detector field of view, and T_{rm} is the room and chopper temperature. For $\varepsilon_{BB} = \varepsilon_{rm} = 1$, $T_{BB} = 500$ K, and $T_{rm} = 300$ K, we may calculate that $L_e^S = 0.1$ W cm^{-2} sr^{-1}.

The value of F_F can be found by adding the normalized rms values of all the Fourier components of the signal waveform in quadrature. For the case presented by the data in Fig. 2-14, clearly only the fundamental sine wave at 1000 Hz is passed by a filter bandwidth of 10 Hz. The second harmonic (2000 Hz) and all higher-order Fourier components are suppressed by the narrow filter. When the source is much smaller than the chopper blade width, the waveform is a square wave for equal blade and gap width (50% duty cycle). The square wave expressed as a Fourier series is

$$A(t) = \frac{4A_0}{\pi} \sum_{n=1}^{\infty} \frac{1}{2n-1} \sin[(2n-1)\omega t] \tag{2.100}$$

where A_0 is the peak amplitude, $\omega = 2\pi f$, and t is time. The fundamental (first) term is

$$A(t) = \frac{4A_0}{\pi} \sin \omega t \tag{2.101}$$

The rms value is

$$a = \left[\frac{16 A_0^2}{\pi^2 T} \int_0^T \sin^2 \omega t \, dt\right]^{1/2} \tag{2.102}$$

where $T = 1/f$ is the period. Then Eq. (2.102) can be written as

$$a = \frac{4 A_0}{\pi \sqrt{2}} = 0.9 A_0 = 0.45(2 A_0) \tag{2.103}$$

where $2 A_0$ is the peak-to-peak amplitude, and $0.45 = F_F$ in this case. The value of F_F for the entire square-wave series (wide bandwidth Δf) is 0.707. When the source aperture is equal to the chopper blade spacing with 50% duty cycle, the signal waveform is triangular. The Fourier series expansion is

$$A(t) = \frac{8 A_0}{\pi^2} \sum_{n=1}^{\infty} \frac{(-1)^{n+1}}{(2n-1)^2} \sin[(2n-1)\omega t] \tag{2.104}$$

and the fundamental is

$$A(t) = \frac{8 A_0}{\pi^2} \sin \omega t \tag{2.105}$$

The rms value of Eq. (2.105) is

$$a = \frac{8 A_0}{\pi^2 \sqrt{2}} = 0.572 A_0 = 0.286(2 A_0) \tag{2.106}$$

It is standard practice to test detectors using a sufficiently narrow bandwidth that only the fundamental term of the signal waveform is measured. The values of F_F therefore range between 0.286 and 0.45. Normal practice produces a signal waveform that is nearly square. Therefore, $F_F = 0.45$ is used to continue this example calculation.

The performance of the photoconductor can now be calculated from the measured data presented in Fig. 2-14. Assuming $\tau = 0.9$ and $F_F = 0.45$ in Eq. (2.98), the rms signal on the detector is about $\phi_e = 5 \times 10^{-9}$ W. This power is spread across the entire blackbody spectrum and the detector, which is characterized by a cutoff wavelength, does not respond to all of this power. By the definition of blackbody voltage responsivity, however,

$$\mathcal{R}_V(500 \text{ K}, 1000 \text{ Hz}) = \frac{30 \times 10^{-3} \text{ V}}{5 \times 10^{-9} \text{ W}} = 6 \times 10^6 \text{ V/W} \tag{2.107}$$

The blackbody NEP is

$$\text{NEP(500 K, 1000 Hz)} = \frac{\phi_e^s}{S/N} = \frac{5 \times 10^{-9}\ \text{W}}{(30 \times 10^{-3}/30 \times 10^{-6})} = 5 \times 10^{-12}\ \text{W}$$

$$(2.108)$$

The blackbody D^* is

$$D^*(500\ \text{K}, 1000\ \text{Hz}) = \frac{\sqrt{A_d \Delta f}}{\text{NEP(500 K, 1000 Hz)}} = \frac{(10^{-2}\ \text{cm}^2)(10\ \text{Hz})}{5 \times 10^{-12}\ \text{W}}$$

$$= 6.3 \times 10^{10}\ \text{cm Hz}^{1/2}\ \text{W}^{-1} \qquad (2.109)$$

An infrared photoconductor that achieves this level of performance must be cooled well below the room temperature of 300 K (see Chapter 4). This fact provides the opportunity to test experimentally whether or not the detector is background limited. A spherical mirror of low emissivity can be placed such that the detector is at the center of curvature. Then only the cold detector itself contributes to the background (except for a very small contribution from the mirror for which $\varepsilon < 0.03$ is commonly achieved). If the noise output is significantly reduced when the mirror blocks out the background, then the detector is operating in a background noise-limited condition. The relation between blackbody D^*_{BLIP} and peak spectral D^*_{BLIP} can then be used to calculate the peak spectral figures of merit. The peak spectral D^*_{BLIP} is calculated from

$$D^*_{\text{BLIP}}(\lambda_P, f) = K(T, \lambda) D^*_{\text{BLIP}}(T_{\text{BB}}, f) \qquad (2.110)$$

Using Fig. 2-12, we obtain

$$D^*(14\ \mu\text{m}, 1000\ \text{Hz}) = 2.3(6.3 \times 10^{10}) = 1.45 \times 10^{11}\ \text{cm Hz}^{1/2}\ \text{W}^{-1}$$

Solving Eq. (2.80) for quantum efficiency,

$$\eta(\lambda_P, f) = \left[D^*_{\text{BLIP}}(\lambda_P, f) \frac{hc \sin \theta_{1/2}}{\lambda_P} \right]^2 \times 2^2 \int_0^{\lambda_P} M_P(\lambda, T)d\lambda \qquad (2.111)$$

Assuming a 20° field of view and $T_B = 300$ K, we obtain

$$\eta(14\ \mu\text{m}, 1000\ \text{Hz}) = 0.59$$

The figures of merit at other than the peak response wavelengths can be found if the spectral response is known. For example, Eq. (2.63) relates the spectral D^* to the spectral voltage responsivity:

$$D^*(\lambda, f) = \frac{\sqrt{A_d \Delta f}}{V_{\text{rms}}} \mathcal{R}_V(\lambda, f) \qquad (2.112)$$

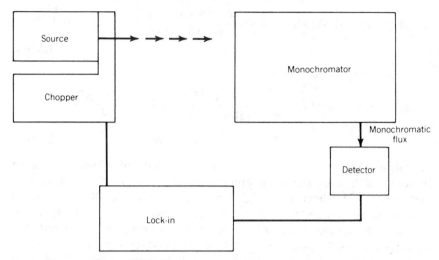

Figure 2-15   Detector spectral response measurement.

The spectral voltage responsivity can be measured by the configuration of Fig. 2-15. The response of the detector under test is ratioed to the calibrated, spectrally flat response of a reference standard thermal detector. The thermal detector will be slower in many cases than the detector under test. This limits the chopping frequency f to low values. The monochromator is an optical device that selects a desired narrow wavelength band of center wavelength λ and width $\Delta\lambda$. The resulting plot displays the spectral responsivity of the detector under test normalized to the responsivity of the reference detector. The spectral response of the detector must be measured if the detector is not BLIP [which means Eq. (2.110) cannot be used], or if the precise spectral characteristics of the detector at other than the response peak must be known.

BIBLIOGRAPHY

Gaskill, J. D., *Linear Systems, Fourier Transforms and Optics*, Wiley-Interscience, New York, 1978.

Grove, A. S., *Physics and Technology of Semiconductor Devices*, Wiley, New York, 1967.

Hooge, F. N., T. G. M. Kleinpenning, and L. K. J. Vandamme, *Reports on Progress in Physics*, **44**(5), 479–532 (1981).

Kittel, C., *Solid State Physics*, Wiley, New York, 1969.

McKelvey, M. C., *Solid State and Semiconductor Physics*, Wiley, New York, 1965.

Nergaard, L. S., and M. Glicksman, *Microwave Solid-state Engineering*, Van Nostrand, New York, 1964.

Pankove, J. I., *Optical Processes in Semiconductors*, Dover, New York, 1971.

Sze, S. M., *Physics of Semiconductor Devices*, 2nd ed., Wiley, New York, 1981.

Weidner, R. T., and R. L. Sells, *Elementary Modern Physics*, 2nd ed., Allyn and Bacon, Boston, MA, 1969.

PROBLEMS

2-1. Justify that the noise equivalent bandwidth is $\Delta f = 1/2\tau$ where $\tau =$ observation time. [Hint: Recall the Fourier transform relationship between time and frequency.]

2-2. The detecting ability of an infrared detector is often described in terms of the specific detectivity D^*, which is defined as follows:

$$D^* \equiv \frac{\sqrt{A\,\Delta f}}{\phi_e}\frac{S}{V_n}$$

where A = detector area,
Δf = noise bandwidth,
S = rms signal voltage,
V_n = rms noise voltage,
ϕ_e = power on detector.

a. Explain why D^* is used, instead of NEP, responsivity, or some other figure of merit.
b. Discuss the use of D^* to characterize a photodetector and a thermal detector.
c. Derive the expression for D^*_{BLIP}, the value of D^* for which the only noise is the photon noise of the background.
d. If the background photon flux becomes very low (i.e., $E_p \approx 0$), what is the corresponding D^* expression if Johnson noise dominates?

2-3. The relationship between blackbody $D^*(T, f)$ and peak spectral $D^*(\lambda_c, f)$ was shown to be (500 K blackbody temperature):

$$KD^*(500, f) = D^*(\lambda_c, f)$$

Graph K versus λ for a blackbody temperature of 2854 K.

2-4.   Calculate D^* (500, 1000) for the following PV detector:

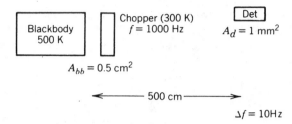

Given the following data: measured S/N = 400, assume atmospheric transmission (T) = 100%, and the detector responds to all wavelengths.

2-5.   Calculate the blackbody detectivity $D^*(500, 1000)$ for the following detector:

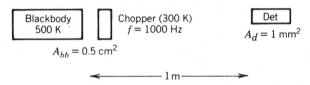

Given the following data: measured S/N = 400, $\Delta f = 1$, assume atmospheric transmission (T) = 100%. This detector has a cutoff wavelength of 14 μm with a 300 K background temperature and a 60° full field of view in the above setup. What is its quantum efficiency (η)?

2-6.   Given a 500 K blackbody and a detector with $\lambda_c = 10$ μm, compute the incident power, ϕ_e, with a 300 K chopper (see 2-5):

a.   A blackbody NEP calculation.

b.   A spectral NEP calculation at $\lambda = 5$ μm, 10 μm $(\Delta\lambda = 1\mu$m).

2-7.   Give a qualitative definition/description (no equations) of NEP and D^*. If you were comparing the manufacturer's specifications for two photoconductive detectors, explain which figure of merit would be most convenient to use. If you were comparing two thermistors (bolometers), explain which figure is most instructive.

2-8.   Compute the Johnson, shot, and $1/f$ noise for a photovoltaic detector with the following parameters: $T = 300$ K, $R_d = 10$ kΩ, $I_{dc} = 10$ mA, $f = 500$ Hz, $\Delta f = 50$ Hz. What is the total detector noise? Which noise dominates? $(\alpha = 2, \beta = 1, a = 1)$

2-9.   Compute the equivalent noise bandwidth of an electronic amplifier assuming a Gaussian power gain $[G^2(f)] = A \exp[-\pi(f - f_0)^2 / \sigma^2]$, $\sigma = 300$ Hz centered at $f = 500$ Hz for the case of:

    a. A white noise source.
    b. A $1/f$ noise source ($\beta = 1$). $f = 1$ lower limit

2-10. Define spectral D star, $[D^*(\lambda, f)]$; blackbody D star $[D^*(T, f)]$, and background-limited peak spectral D star $[D^*_{BLIP}(\lambda_P, f)]$. Discuss, in particular, the radiometric normalization considerations in calculating or measuring the above quantities.

2-11. Discuss the various modes of operation for a photovoltaic detector. List advantages and disadvantages for each mode.

2-12. What determines the spectral response of a detector?

2-13. Identify those characteristics that distinguish the following photodiodes:

    a. *P-N.*
    b. *P-I-N.*
    c. Tunnel.
    d. Avalanche.

2-14. As an optical sensor system designer (optics, detector, electronics) what would you want to know about the detector?

2-15. How would you derive the Johnson noise-limited D^* for a photodetector?

2-16. Using mathematical expressions, relate responsivity ($\mathcal{R}$), noise equivalent power (NEP), and D star (D^*). Why is more than one figure of merit needed to specify a detector?

2-17. Explain noise equivalent bandwidth in relation to noise sources. In addition, discuss the statement that "all noises vary as the square root of the bandwidth."

2-18. Why is power generated only in the fourth quadrant of a photovoltaic detector?

2-19. Show that for every 11°C change in semiconductor temperature the carrier concentration doubles. (silicon @300).

2-20. Verify the Fourier transform relationship between the time voltage responsivity expression and the frequency domain expression for thermal detectors.

2-21. Calculate the Johnson noise voltage across a parallel combination of resistors as shown ($\Delta f = 100$ Hz).

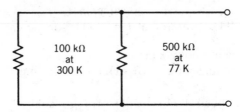

2-22. Calculate the rms power from radiation which is chopped in a square wave.

CHAPTER 3

PHOTOVOLTAIC DETECTION THEORY

3-1 INTRINSIC PHOTOVOLTAIC EFFECT

The photovoltaic effect requires a potential barrier with an electric field in order to operate. A semiconductor $P-N$ junction is usually employed to obtain these conditions. Figure 3-1 illustrates the energy-band diagram of a $P-N$ junction. An incident photon of energy greater than or equal to the energy gap E_g can create a hole–electron pair as shown in Fig. 3-1. The electric field of the junction will not allow the hole–electron pair to recombine. The photoelectrons are therefore available to produce a current through an external circuit.

Two basic requirements for photovoltaic operation are that the incident photons have sufficient energy to excite photoelectrons across the energy gap and that the temperature of the detector be low enough so that the electrons are not thermally excited across the band gap. The photon restriction is

$$h\nu = \frac{hc}{\lambda} \geqq E_g \qquad (3.1)$$

where h = Planck's constant = 6.626×10^{-34} J sec,
ν = the optical radiation (light) frequency in Hz,
c = the speed of light = 2.998×10^{10} cm sec^{-1},
λ = the optical radiation wavelength in cm,
E_g = the energy gap in J.

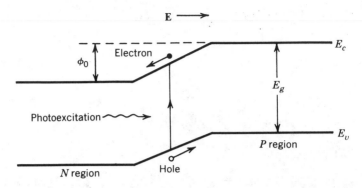

Figure 3-1 Photovoltaic effect.

This implies an upper limit to the wavelength at which the photovoltaic effect will operate:

$$\lambda_{max} = \frac{1.24}{E_g} \qquad \frac{1.24}{0.22} \qquad \frac{1.24}{1.12} \qquad (3.2)$$

[handwritten annotations: InSb MCT Si ; λ_{max} 5.636   1.107]

where E_g = the junction energy gap in electron volts (eV),
$\quad\lambda$ = wavelength in μm.

The thermal constraint is

$$\frac{kT}{q} \ll E_g \qquad (3.3)$$

[handwritten: $T \ll \left(\frac{q}{k}\right) E_g$; $\frac{q}{k} = \frac{1.6E\text{-}19}{1.38E\text{-}23} \sim 10^4$]

where $k = 1.38 \times 10^{-23}$ W sec K^{-1} is Boltzmann's constant,
$\quad T$ = the temperature (K).

This ensures that electrons will be available for photoexcitation. Table 3-1 lists some values of E_g, λ_{max}, and operating temperature for common photovoltaic devices.

The absorption of photons to produce hole–electron pairs only takes place in the P–N junction area. The effective width of the P–N junction in centimeters (W_0) is

$$W_0 = 10^3 \sqrt{\frac{K\phi_0(N_d + P_a)}{N_d P_a}} \qquad (3.4)$$

where K is the dielectric constant (e.g., $K \cong 16$ for Ge, $K \cong 12$ for Si), N_d is the concentration (number per cubic centimeter) of donors in the N

Table 3-1

Detector	E_g (eV)	T (typical operation) (K)	λ_{max} (10^{-4} cm)
InSb	0.22	77	5.5
PbS	0.42	193	3
Ge	0.67	193	1.9
Si	1.12	300	1.1
CdSe	1.8	300	0.69
CdS	2.4	300	0.52

region, P_a is the concentration (cm^{-3}) of acceptors in the P region, and ϕ_0 is the potential barrier:

$$\phi_0 \approx \frac{kT}{q} \ln\left(\frac{N_p P_n}{N_i^2}\right) \tag{3.5}$$

where N_p is the concentration of electrons in the P region ($N_p \approx P_a$), P_n is the concentration of holes in the N region ($P_n \approx N_d$), and N_i is the intrinsic carrier concentration. The intrinsic carrier concentration varies as the cube of the absolute temperature. Therefore, as the junction temperature decreases, the potential barrier increases and so does the width of the junction. Figure 3-2 illustrates that the effective width of the junction is not symmetrical about the limiting position of the junction as W_0 approaches zero width. This effect is due to the mobility differences between current carriers in the two regions:

$$L_h = \sqrt{\frac{kT\mu_h \tau_h}{q}} \tag{3.6}$$

and

$$L_e = \sqrt{\frac{kT\mu_e \tau_e}{q}} \tag{3.7}$$

where L_h and L_e are the hole and electron diffusion lengths, μ_h and μ_e are the hole and electron mobilities, and τ_h and τ_e are the hole and electron lifetimes, respectively.

The structure of the P–N junction sensitivity distribution described above does not imply that photovoltaic detectors must vary in sensitivity

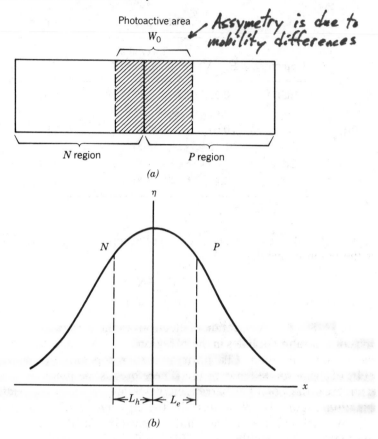

Figure 3-2   (*a*) Effective width of *P–N* junction (*W₀*). (*b*) Quantum efficiency of converting photons to photoelectrons decreases with distance from the junction.

across the surface. Figure 3-3 illustrates that it is possible to orient the junction such that all incident photons must traverse the entire width W_0 of the junction twice, unless they are absorbed first. The reflective coating on the bottom of the detector provides this "double jeopardy" for the photons as well as serving to make electrical contact to the semiconductor. If the carrier diffusion length is long compared to the distance between electrical contacts, and if the material is essentially transparent to the photons of interest, then a quantum efficiency approaching twice the peak value of Fig. 3-2*b* may be approached in many cases.

The photovoltaic effect, then, is the generation of a voltage difference by incident photons. The voltage arises because the *P–N* junction electric

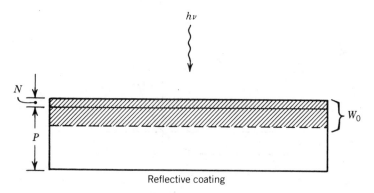

Figure 3-3 Detector configuration for which all incident photons traverse the P–N junction.

field forces photoelectrons to remain in the N region, and the corresponding holes produced by the photoexcitation are forced to occupy the P region. The maximum possible voltage produced in this way is labeled ϕ_0 in Fig. 3-1. The actual voltage produced is a function of the load that is connected to the photovoltaic device, and the properties of the semiconducting material used.

3-2   PHOTOCURRENT

The continuity equation for a P–N junction determines the current through the junction. The form of the continuity equation presented below states that the rate of change of the concentration of holes (electron acceptors) above the thermally generated concentration is equal to the rate of new hole generation less the rate at which the holes disappear through recombination:

$$\frac{d(\Delta P)}{dt} = \left[g + \frac{D_e}{q} \frac{\partial^2(\Delta P)}{\partial x^2} \right] - \frac{\Delta P}{\tau_h} \tag{3.8}$$

where ΔP is the number of holes per cm^3 above the thermal equilibrium value, D_e is the electron diffusion constant $(kT\mu_e/q)$, g is the rate of hole generation by nonthermal processes (e.g., photon-generated hole–electron pairs), and τ_h is the hole lifetime. For the boundary conditions of steady-state current $[d(\Delta P)/dt = 0]$ and no photocurrent $(g = 0)$, the

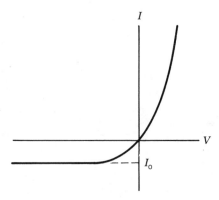

Figure 3-4 *P–N* junction diode curve.

continuity equation can be solved for the hole current:

$$I_h = \frac{qN_pD_e}{L_e}(e^{qV/kT} - 1) \tag{3.9}$$

where N_p is the concentration of electrons in the P region, L_e is the electron diffusion length ($\sqrt{D_e\tau_e}$), and V is the externally applied voltage. The electron continuity equation can be similarly solved for the electron current and a total steady-state dark current can be found:

$$I_{\text{dark}} = q\left(\frac{N_pD_e}{L_e} + \frac{P_nD_n}{L_n}\right)(e^{qV/kT} - 1) \tag{3.10}$$

The current characteristics of the $P–N$ junction in the absence of light are summarized in Fig. 3-4. The dark current as a function of applied voltage is

$$I_{\text{dark}} = I_0[e^{qV/kT} - 1] \tag{3.11}$$

where q is the charge on an electron $= 1.6 \times 10^{-19}$ C and I_0 is the reverse saturation current characteristic of the diode. The reverse saturation current is a strong function of temperature for a given junction. In silicon $P–N$ junctions, for example, I_0 approximately doubles for each temperature increase of 10 K.

Exposing the $P–N$ junction to optical radiation ($g > 0$ in the continuity equation) shifts the $I(V)$ curve downward by an amount known as the photocurrent (photon-generated current), I_p. This shift is shown in Fig. 3-5. The photocurrent is given by

$$I_p = \eta qE_eA_d\frac{\lambda}{hc} \tag{3.12}$$

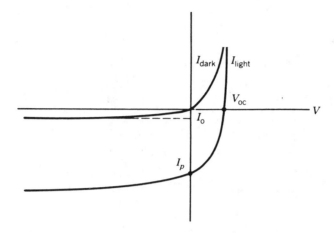

Figure 3-5 Diode curve shift during exposure to light.

where η is the quantum efficiency (the number of photoelectrons generated per incident photon in the photovoltaic case), E_e is the incident radiant flux density (irradiance) in W cm^{-2}, A_d is the area of the photovoltaic device, c is the speed of light, and $h = 6.6 \times 10^{-34}$ W sec^2. This short-circuit current is a linear function of incident flux typically over a range of 10^7. The total current in the presence of light is therefore

$$I_{light} = I_0[e^{qV/kT} - 1] - \eta q E_e A_d \frac{\lambda}{hc} \qquad (3.13)$$

This section has introduced the photovoltaic device as a current generator, but it is also commonly used as a power generator. The short-circuit current delivered by the device is given by Eq. (3.12). The open-circuit voltage is

$$V_{oc} = \frac{kT}{q} \ln\left[\frac{I_p + I_0}{I_0}\right] \qquad (3.14)$$

The device can provide power by operating into a finite load resistance R_L as shown in Fig. 3-6. This results in a power generated, represented by the shaded area in the figure. This power is the product of some VI, where $0 < V < V_{oc}$ and $0 < I < I_p$. Power generation does not fall within the scope of this book. A simple theoretical model for predicting solar cell performance is included in Appendix B so that the interested reader can pursue the subject further.

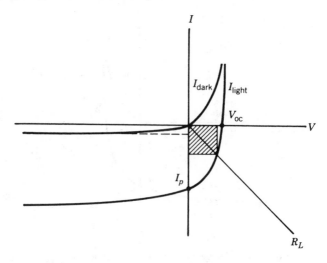

Figure 3-6   Photovoltaic power generation.

3-3   RESPONSIVITY AND QUANTUM EFFICIENCY

Photovoltaic devices are commonly operated as current generators with a current responsivity expressed as the current in amperes divided by the incident power in watts required to generate that current. Dividing the photocurrent of Eq. (3.12) by the incident power Φ_e yields the current responsivity $\mathscr{R}_i$:

$$\mathscr{R}_i = \frac{I_p}{\Phi_e} = \frac{\eta q \lambda}{hc} \tag{3.15}$$

Substituting for the constants we have

$$\mathscr{R}_i = 0.808 \, \eta \lambda \text{ A W}^{-1} \tag{3.16}$$

where λ is in micrometers. It is seen from Eq. (3.16) that the response of an ideal photovoltaic device is proportional to wavelength (for constant quantum efficiency) until the maximum wavelength of Eq. (3.2) is reached (see Fig. 3-7).

Confusion often arises over "why" the current responsivity increases with wavelength. The source of the confusion is that many people will interpret Fig. 3-7 as an indication that the detector physically "works better" at longer wavelengths. Actually, this is an artifact of the units being used. The common units for current responsivity of amperes per

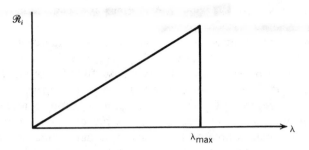

Figure 3-7   Ideal responsivity for constant quantum efficiency.

watt are specified for convenience because the power incident on a detector is easily measured. Suppose, instead, current responsivity was defined in units of amperes per photon. The expression for photocurrent as a function of photon irradiance E_p photons per second per centimeter squared is

$$I_p = \eta q E_p A_d \tag{3.17}$$

The photon current responsivity is then

$$\mathcal{R}_{ip} = \frac{I_p}{\Phi_p} = \eta q \tag{3.18}$$

In this case, for constant quantum efficiency, the responsivity is constant until the cutoff wavelength is reached. In practice, the quantum efficiency is found to rolloff at both the long-wavelength and short-wavelength regions. The short-wavelength rolloff is due to photon absorption at the surface of the detector. Because of surface traps and recombination times, the charge is lost before it gets to the $P-N$ junction. The long-wavelength cutoff (λ_c) is fixed by the energy gap, but depending on the detector bias the various curves shown in Fig. 3-8 can be produced. This is often called

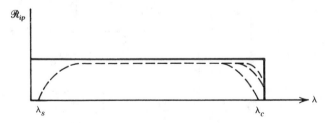

Figure 3-8   Ideal photon responsivity.

extended red response. By increasing the bias, the recombination rate goes down as the carriers are swept out.

The loss in quantum efficiency from the ideal value of 1 is due to (1) optical reflection losses (Fresnel); (2) surface traps or recombination centers; (3) absorption coefficient varying as a function of wavelength; (4) photogenerated carriers being created further than a diffusion length from the depletion region. Physical insight into photon detectors is best obtained in photon rather than energy or power units. Convenience in calculating the measurable performance of actual devices is enhanced by using the common definition of $\mathcal{R}_i$ in Eq. (3.15). The voltage responsivity can be similarly defined (when $I_p \ll I_o$):

$$\mathcal{R}_V = \mathcal{R}_i R_d = \frac{\eta R_d q\lambda}{hc} = 0.808\,\eta\lambda R_d \tag{3.19}$$

where R_d is the detector resistance.

The definition of current responsivity has implicitly assumed that the detector is working into a short circuit, because I_p is the short-circuit current. Amplification of the short-circuit current is commonly performed using a transimpedance amplifier circuit (Fig. 3-9). The characteristics of an operational amplifier are such that the detector sees a virtual short circuit to ground in Fig. 3-9. There is no voltage across the detector, so the output voltage must be

$$V_{\text{out}} = I_p R_f \tag{3.20}$$

Preamplifier performance will be treated in more detail in Chapter 4 of this book.

Current responsivity for a given detector is easily measured by irradiating the detector with a known signal power. We then have $\mathcal{R}_i = I_p/\Phi_e =$

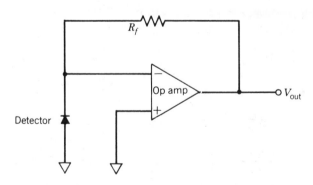

Figure 3-9   Transimpedance amplifier circuit.

$V_{out}/\Phi_e R_f$. This also constitutes a measurement of the quantum efficiency of the detector at one wavelength if the irradiating signal were monochromatic:

$$\eta = \frac{\mathscr{R}_i}{0.808\lambda} \tag{3.21}$$

by rearranging Eq. (3.16). If the source were not monochromatic, the average quantum efficiency over the wavelength interval emitted can be obtained from Eq. (3.21) if the signal were of constant irradiance with wavelength. The term quantum efficiency is itself a monochromatic concept, however.

3-4 FREQUENCY RESPONSE

A $P-N$ junction photovoltaic device has a frequency response determined by the characteristics of the device itself and by the circuit in which it operates. The detector and its circuit load have an effective overall resistance R and capacitance C, which implies a cutoff frequency:

$$f_c = \frac{1}{2\pi RC} \tag{3.22}$$

where f_c is the frequency at which the response drops by three decibels (see Fig. 3-10). The use of f to denote frequency in electrical circuits is traditional in order to prevent confusion with ν as the symbol for the frequency of optical radiation (electromagnetic radiation).

An equivalent circuit of a detector and its electronic load R_L is shown

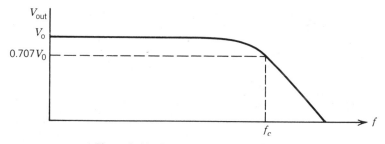

Figure 3-10 Detector frequency response.

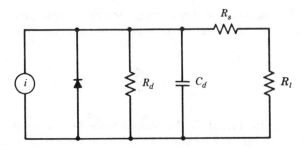

Figure 3-11   Detector equivalent circuit.

in Fig. 3-11. The overall resistance is

$$R_T = \cfrac{1}{\cfrac{1}{R_d} + \cfrac{1}{R_s + R_L}}$$

(3.23)

and the cutoff frequency is therefore

$$f_c = \frac{1}{2\pi R_T C_d}$$

(3.24)

We will consider a typical silicon photovoltaic detector as a specific example. It is common for silicon detectors working into a short circuit (Fig. 3-9) to have a capacitance of $0.01\ \mu\mathrm{F\ cm}^{-2}$. For a detector area of $1\ \mathrm{mm}^2$, $C_d = 10^{-10}\ \mathrm{F}$. Assuming $R_d = 10^9\ \Omega$ and $R_s = 0.01\ \Omega$, if $R_L = 50\ \Omega$, then, essentially, $R_T = R_L = 50\ \Omega$ from Eq. (3.23). Equation (3.24) now yields $f_c = 30\ \mathrm{MHz}$.

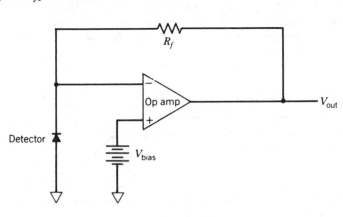

Figure 3-12   Detector bias circuit.

It is possible to decrease the capacitance of the silicon junction by applying a reverse bias (see Fig. 3-12). The reverse bias causes an increase in ϕ_0 and therefore an increase in the depletion layer width W_0 of Eq. (3.4) (separation of the P and N regions) and a decrease in C_d. For the silicon detector in our example, a 100 V reverse bias would be expected to decrease C_d to 10^{-11} F and therefore increase f_c to 300 MHz. The reverse bias does increase the electrical noise in the detector, and a design trade-off between noise and frequency response would have to be made in practice. C_d can also be decreased by placing an intrinsic layer in the junction (*PIN* diode). For more details on the frequency response of silicon diodes, see Willardson and Beer (1977).

3-5 NOISE SOURCES

Noise is the random fluctuation of the electrical output signal (voltage, current, or power). The assumption that noise is random is justified because any predictable unwanted signal can be cancelled by a signal of opposite polarity and equal magnitude. Because noise cannot be cancelled in this way, it fundamentally limits the achievable performance of a given detection system.

There are three classes of noise to be considered. This section first considers sources of noise generated within a photovoltaic detector. These sources are Johnson noise and $1/f$ noise. Noise due to the fluctuation of the incident optical radiation is called photon noise or quantum noise. The third source of noise is in the associated preamplifier electronics.

The Johnson noise (Nyquist noise, thermal noise) is caused by the thermal motions of charged particles in a resistor. The root-mean-square (rms) electrical noise power fluctuations per hertz of bandwidth is just due to the thermal energy:

$$P_{rms} = kT\Delta f \qquad (3.25)$$

where P_{rms} = noise power in watts,
$\quad k$ = Boltzmann's constant = 1.38×10^{-23} W sec K^{-1},
$\quad T$ = resistor temperature in degrees kelvin,
$\quad \Delta f$ = electrical bandwidth in hertz.

The noise current and voltage for a given resistance R in ohms are therefore (as derived in Chapter 2):

$$i_{rms} = \sqrt{\frac{4kT\Delta f}{R}} \qquad (3.26)$$

$$V_{rms} = \sqrt{4kTR\Delta f} \tag{3.27}$$

where i_{rms} = noise current in amperes,

V_{rms} = noise voltage in volts.

There exists a noise source whose amplitude decreases approximately as $1/f$, where f is the frequency at which the noise is measured. In spite of extensive theoretical efforts to explain the origin of this "one over f" spectral shape, no generally accepted mechanism for the production of this noise has been proposed. We therefore present the experimentally measured relation:

$$i_{rms} = \sqrt{\frac{BI_{DC}^{\alpha}\Delta f}{f^{\beta}}} \tag{3.28}$$

where B = constant dependent on the particular detector used,

I_{DC} = direct current through the detector (i.e., AC current does not contribute to $1/f$ noise),

α = typically 2 but values have been measured in the range 1.5–4,

β = typically 1 but values range between 0.8 and 1.5.

Because this noise is a function only of the direct-current component of the detector current, it is possible to essentially eliminate this noise for photovoltaic detectors. By replacing the fixed bias supply of Fig. 3-12 with a variable bias, it is possible to adjust I_{DC} to zero. The alternating current produced by most signals of interest will not contribute to $1/f$ noise.

A fundamental limit to the achievable performance of any detection system is noise present in the incident optical radiation. It is a sufficiently accurate assumption for virtually all applications at optical frequencies to assume that the photon arrival rate has a Poisson distribution about the mean number of photons arriving per time interval, $\bar{N}$:

$$P(N) = \frac{\bar{N}^N e^{-\bar{N}}}{N!} \tag{3.29}$$

where $P(N)$ is the probability of N photons arriving within one time interval. For large numbers of photons the Poisson distribution approaches the shape of a normal or Gaussian distribution. For any number of photons N, the standard deviation is

$$\sigma = \sqrt{\bar{N}} \tag{3.30}$$

The physical interpretation of noise is simply that the noise level is an error of magnitude σ in the accuracy to which the signal level can be measured. The uncertainty in the arrival rate of incident photons implies that the signal-to-noise ratio (S/N) for a photon stream is

$$S/N = \frac{\bar{N}}{\sigma} = \frac{\bar{N}}{\sqrt{\bar{N}}} = \sqrt{\bar{N}} \tag{3.31}$$

Therefore, a noiseless detector with unit quantum efficiency followed by noiseless amplifying electronics will have an output signal-to-noise ratio of $\sqrt{\bar{N}}$. The photon noise present in the incident optical radiation itself represents the fundamental limit to detection system performance. A detector that has a noise level low enough for background photon noise to limit performance is said to have achieved BLIP (background-limited photodetector) performance.

The expression for the electrical output noise due to the optical signal can be derived from the expression for shot noise in a diode:

$$i_{\text{rms}} = \sqrt{2qI\Delta f} \tag{3.32}$$

where I is the total current through the diode stated previously:

$$I = I_0[e^{qV/kT} - 1] - \eta q E_e A_d \frac{\lambda}{hc} \tag{3.33}$$

Because photovoltaic detectors are usually operated into a virtual short circuit (Fig. 3-9), $V = 0$, and we have

$$I = \eta q E_e A_d \frac{\lambda}{hc} = q\eta E_p A_d \tag{3.34}$$

Substituting Eq. (3.34) for I in Eq. (3.32),

$$i_{\text{rms}} = \sqrt{2q^2 \eta E_p A_d \Delta f} \tag{3.35}$$

The photon irradiance E_p is the sum of the irradiance due to the desired signal E_p^S and the irradiance due to unwanted background radiation E_p^B. The electrical noise current and voltage due to incident photons are therefore:

$$i_{\text{rms}} = \sqrt{2q^2 \eta [E_p^S + E_p^B] A_d \Delta f} \tag{3.36}$$

$$V_{\text{rms}} = \sqrt{2q^2 \eta [E_p^S + E_p^B] A_d R_d^2 \Delta f} \tag{3.37}$$

Noise powers from different sources can be added directly to obtain the total resulting noise power:

$$P_{T\,\text{rms}} = P_{1\,\text{rms}} + P_{2\,\text{rms}} + P_{3\,\text{rms}} + P_{3\,\text{rms}} + \cdots \qquad (3.38)$$

Therefore, noise currents and voltages must be added in quadrature:

$$i_{T\,\text{rms}} = \sqrt{i_{1\,\text{rms}}^2 + i_{2\,\text{rms}}^2 + i_{3\,\text{rms}}^2 + \cdots} \qquad (3.39)$$

$$V_{T\,\text{rms}} = \sqrt{V_{1\,\text{rms}}^2 + V_{2\,\text{rms}}^2 + V_{3\,\text{rms}}^2 + \cdots} \qquad (3.40)$$

Adding the noise voltages due to Johnson noise, $1/f$ noise, and photon noise:

$$V_{T\,\text{rms}} = \sqrt{4kTR_d\Delta f + BI_{\text{DC}}^\alpha f^{-\beta} R_d^2 \Delta f + 2q^2 \eta [E_p^S + E_p^B] A_d R_d^2 \Delta f} \qquad (3.41)$$

This represents the total noise voltage present across a photovoltaic detector.

3-6 FIGURES OF MERIT

The simple concept of responsivity $\mathscr{R}$ introduced in Section 3.3 needs to be examined in more detail. As treated in Section 3.3, spectral responsivity $\mathscr{R}_i(\lambda, f)$ was discussed:

$$\mathscr{R}_i(\lambda, f) = \frac{I_p}{\Phi_e(\lambda, f)} \qquad (3.42)$$

where $\Phi_e(\lambda, f)$ is assumed to be a monochromatic incident power at wavelength λ, modulated at frequency f. A quite different responsivity $\mathscr{R}_i(T, f)$ is often quoted:

$$\mathscr{R}_i(T, f) = \frac{I_p}{\Phi_e^{\text{BB}}(T, f)} \qquad (3.43)$$

where $\Phi_e^{\text{BB}}(T, f)$ is the total power incident on a detector due to a blackbody source at temperature T, modulated by a frequency f:

$$\Phi_e^{\text{BB}}(T) = A_d \int_0^\infty E_e(\lambda, T)d\lambda \qquad (3.44)$$

where $E_e(\lambda, T)$ is the blackbody irradiance. Note that $\mathscr{R}_i(T, f)$ is defined as the ratio of the photocurrent to the incident optical radiation integrated over all wavelengths, even though the detector does not respond to wavelengths above λ_{max} defined by Eq. (3.2). This responsivity $\mathscr{R}_i(T, f)$ is known as the blackbody responsivity.

The minimum signal power for which a usable signal-to-noise ratio is produced by the detector is an important figure of merit. The noise equivalent power (NEP) is defined as the root-mean-square incident signal power on a detector that produces an rms signal-to-noise ratio of 1. If the noise current i_{rms} is known, the NEP may be found by solving for Φ_e at which S/N = 1:

$$\text{S/N} = \frac{I_p}{i_{rms}} = \frac{\Phi_e(\lambda/hc)q\eta}{i_{rms}} = 1 \tag{3.45}$$

solving for $\Phi_e = \text{NEP}$:

$$\text{NEP}(\lambda, f) = \frac{hci_{rms}}{q\eta\lambda} \tag{3.46}$$

If the incident power is due to blackbody radiation integrated over all wavelengths, NEP(T, f), the blackbody noise equivalent power, is defined. The spectral NEP(λ, f) is a monochromatic quantity defined only at wavelength λ. A low NEP implies that a small signal can be detected. Therefore, the lower the NEP, the more sensitive the detector is. The NEP is a function of the parameters under which it is measured, such as the area of the detector A_d and the noise bandwidth Δf.

A figure of merit that is normalized with respect to the detector area and noise bandwidth allows comparisons between detectors:

$$D^*(\lambda, f) = \frac{\sqrt{A_d\Delta f}}{\text{NEP}(\lambda, f)} \tag{3.47}$$

D^* is read "dee star" and is also frequently called detectivity. Note that the units of D^* are cm Hz$^{1/2}$ W^{-1}. Alternate expressions to (3.47) include

$$D^* = \frac{\sqrt{A_d\Delta f}}{\text{NEP}} = \frac{\sqrt{A_d\Delta f}}{V_{rms}}\mathscr{R}_V = \frac{\sqrt{A_d\Delta f}}{\Phi_e}\text{S/N} \tag{3.48}$$

An interpretation of the meaning of D^* suggested by Eq. (3.48) is that it is the signal-to-noise ratio resulting when 1 W of optical radiation is incident on a detector of area 1 cm^2 measured with a noise bandwidth of 1 Hz. If the peak spectral D^*, $D^*(\lambda_{max}, f)$, is known, the $D^*(\lambda, f)$ for $\lambda \leq \lambda_{max}$ can be estimated from

$$D^*(\lambda, f) = \frac{\lambda}{\lambda_{max}}D^*(\lambda_{max}, f) \tag{3.49}$$

where $D^*(\lambda, f) = 0$ for $\lambda > \lambda_{max}$. It is even possible to calculate the blackbody $D^*(T. f)$ from the peak spectral D^*:

$$D^*(T, f) = \frac{\int_0^{\lambda_{max}} (\lambda / \lambda_{max}) D^*(\lambda_{max}, f) E_e^{BB}(T, \lambda) d\lambda}{\int_0^{\infty} E_e^{BB}(T, \lambda) d\lambda} \qquad (3.50)$$

It is now possible to examine how some of the detector parameters affect performance. First, an expression for $D^*(\lambda, f)$ containing all the noise sources explicitly is developed. Then various limits to detector performance will be explored.

An expression for all the noise sources involved in typical photovoltaic detector operation must include the detector noise sources contained in Eq. (3.41), as well as the Johnson noise of the feedback resistor R_f and the noise generated in the preamplifier electronics of Fig. 3-9:

$$V_{T\,rms} = [2q^2 \eta (E_p^S + E_p^B) R_d^2 A_d \Delta f + 4kT_d R_d \Delta f + 4kT_f R_f \Delta f$$
$$+ BI_{DC}^{\alpha} R_d^2 \Delta f f^{-\beta} + V_{preamp}^2]^{1/2} \qquad (3.51)$$

where all noise magnitudes are related to the equivalent preamplifier input.

Combining Eq. (3.51) and the responsivity expression

$$\mathcal{R}_V = \frac{qR_d\eta\lambda}{hc} \qquad (3.52)$$

with the D^* definition

$$D^*(\lambda, f) = \frac{\sqrt{A_d\Delta f}}{V_{rms}} \mathcal{R}_V \qquad (3.53)$$

we obtain the desired expression:

$$D^*(\lambda, f) = \frac{\lambda}{hc} \sqrt{\frac{\eta}{2E_p + \dfrac{4kT_d}{\eta q^2 R_d A_d} + \dfrac{4kT_f R_f}{\eta q^2 R_d^2 A_d} + \dfrac{BI_{DC}^{\alpha} f^{-\beta}}{\eta q^2 A_d} + \dfrac{V_{preamp}^2}{\eta q^2 R_d^2 A_d \Delta f}}}$$

$$(3.54)$$

where $\lambda \leq \lambda_{max}$.

It was stated in Section 3.5 that $1/f$ noise can essentially be eliminated in high-performance photovoltaic detectors. In many cases sufficiently low noise preamplifiers can be obtained so that the contribution to overall system noise is very small. We may then assume

$$D^*(\lambda, f) \approx \frac{\lambda}{hc} \sqrt{\frac{\eta}{2E_p + \dfrac{4kT_d}{\eta q^2 R_d A_d} + \dfrac{4kT_f R_f}{\eta q^2 R_d^2 A_d}}} \qquad (3.55)$$

The ultimate performance limit for a photovoltaic detector is the noise due to the photon irradiance E_p. This noise dominates in Eq. (3.55) whenever

$$E_p \gg \frac{2k}{\eta q^2 R_d A_d} \left[T_d + T_f \frac{R_f}{R_d} \right] \qquad (3.56)$$

This condition may exist if the detector and feedback resistor temperature are low, the quantum efficiency is high, and the "RA product," $R_d A_d$, is high. The RA product is considered to be one term, because if A_d is increased, R_d decreases in general. When these conditions are met [inequality (3.56) holds], the detector may achieve background-limited performance:

$$D^*_{\text{BLIP}}(\lambda, f) = \frac{\lambda}{hc} \sqrt{\frac{\eta}{2E_p}} \qquad (3.57)$$

Note that D^*_{BLIP} becomes large without limit as E_p decreases toward zero. In cases of very low background irradiance, achieving D^*_{BLIP} becomes impractical. In space or astronomical applications for which E_p is very low, state-of-the-art detectors are Johnson noise limited:

$$D^*_{\text{JOLI}}(\lambda, f) = \frac{\lambda}{hc} \frac{\eta q}{2} \sqrt{\frac{R_d A_d}{kT_d + kT_f (R_f/R_d)}} \qquad (3.58)$$

where D^*_{JOLI} is Johnson-noise-limited detectivity. It is usually possible to obtain detector Johnson-noise-limited performance by a suitable choice of T_f and R_f. In this case,

$$D^*_{\text{JOLI}}(\lambda, f) = \frac{\lambda \eta q}{2hc} \sqrt{\frac{R_d A_d}{kT_d}} = \frac{\mathscr{R}_i}{2} \sqrt{\frac{R_d A_d}{kT_d}} \qquad (3.59)$$

Note that for either BLIP or JOLI operation it is desirable to have a large $R_d A_d$ product.

BIBLIOGRAPHY

Keyes, R. J., *Optical and Infrared Detectors, Vol. 19, Topics in Applied Physics*, Springer-Verlag, New York, 1980.

Kingston, R. H., *Detection of Optical and Infrared Radiation*, Springer-Verlag, New York, 1978.

Kruse, P. W., L. D. McGlaughlin, and R. B. McQuistan, *Elements of Infrared Technology*, Wiley, New York, 1963.

Sze, S. M., *Physics of Semiconductor Devices*, 2nd ed., Wiley-Interscience, New York, 1981.

Willardson, R. K., and A. C. Beer, *Semiconductors and Semimetals*, Academic Press, New York, 1977.

Wolfe, W. L. and G. J. Zissis, *The Infrared Handbook*, Superintendent of Documents, Washington, D.C., 1979.

PROBLEMS

3-1.   Calculate the noise at the output of a photovoltaic detector connected to an operational amplifier as shown below:

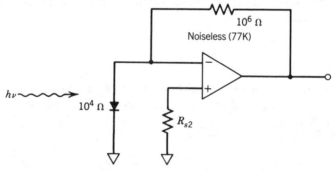

The pre-amp voltage noise generated is equal to 30 nV (3×10^{-8}V) rms and the current noise generated is 10^{-13} A rms. The photon flux (E_p) is 10^{16} q sec^{-1} cm^{-2}. Assume $\eta = 0.5$ and $\Delta f = 1$, $A_d = 1$ mm^2. Which is the dominant noise? (T = 300 K)

3-2.   Explain the significance of capacitance effects in a photovoltaic detector in a typical operational amplifier circuit.

3-3.   How can the spectral response of a silicon photovoltaic be tailored to different spectral responsivity peaks?

3-4.   Given a photovoltaic detector with a resistance less than the optimum source resistance for the preamplifier, show that placing a resistor in series with the detector such that the total resistance equals the desired value is not appropriate.

3-5. List the factors that increase the temperature of a cooled detector during operation.

3-6. A photodiode (InSb) is operated at 77 K (liquid nitrogen). The detector has a resistance of 10 MΩ and is used in a preamplifier circuit with a feedback resistor of $10^8 \Omega$ @77 K. If 1 nA of photogenerated current is flowing through the detector, what is the total voltage noise output measured in a bandwidth of 1 Hz? (see problem 3-1).

3-7. Plot the BLIP peak spectral $D^*(\lambda, f)$ versus λ (1–100 μm) of a photovoltaic detector for background temperatures of:
   a. 28 K (Liquid neon).
   b. 77 K (Liquid nitrogen).
   c. 197 K (dry ice and acetone).

3-8. The expression for the current from a photovoltaic detector is $I_0(e^{qV/\beta kT} - 1) - I_g$, where I_g is the photogenerated current and β is found empirically. If two photodetectors were exactly the same, except $\beta = 1$ for one and $\beta = 2$ for the other, which detector would you choose to use? Why? What operating condition could be varied so that a single photodiode could give these two situations?

3-9. Given a Si avalanche photodiode operating at 300 K:
   a. Compute the minimum gain M necessary to ensure shot-noise-limited operation. ($R_L = R_d = 10^9 \, \Omega$, $\phi_e = 10^{-9}$ W, $\lambda = 800$ nm, $\eta = 0.85$, $V = 100$ volts, $P = 2.5$).
   b. What is the S/N if $M = 100$ ($\Delta f = 1$)?

3-10. Rederive the total noise current expression (shot and Johnson) for a photovoltaic detector for the case of $\beta \neq 1$. Derive an expression for the bias voltage as a function of β which minimizes the noise and plot V_B versus β ($0.5 \leq \beta \leq 3$).

3-11. Derive the spectral NEP expression for BLIP operation of a photovoltaic detector.

3-12. Calculate the spectral NEP for a Johnson noise limited photovoltaic detector. Can you think of a case when the detector is likely to be Johnson noise limited?

3-13. Discuss the operation of a classical photovoltaic in all four quadrants of the $V-I$ plot. In addition, discuss with appropriate equations operation on the voltage and current axes.

3-14. A photodiode is operated at 77 K (liquid nitrogen). The detector has a resistance of 10 MΩ:
   a. What is its Johnson noise voltage ($\Delta f = 1$)?

    b.   If 5 nA is flowing through the detector, what is the total output voltage due to Johnson and shot noise sources?

3-15.   Plot peak spectral D star, $D^*(\lambda_c, f)$, versus wavelength ($1 < \lambda_c < 100\ \mu$m) for a 2π sr, 295 K background for a photovoltaic detector.

3-16.   List the factors that cause the quantum efficiency to be less than 1 in a photodetector.

3-17.   For a silicon photovoltaic with a $V-I$ characteristic as shown below, calculate the blackbody responsivity for the following operating conditions:

    a. $R_L = 0$.

    b. $R_L = \infty$.

    c.   What is the dynamic resistance of this detector at zero bias?

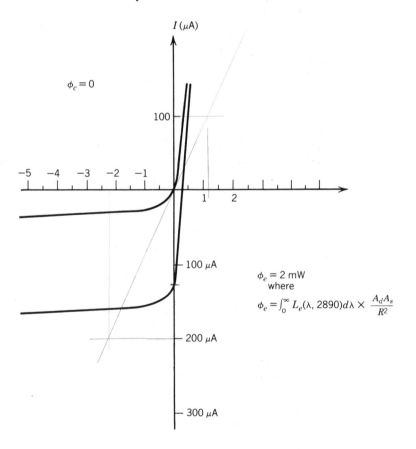

$\phi_e = 2$ mW

where

$$\phi_e = \int_0^\infty L_e(\lambda, 2890)d\lambda \times \frac{A_d A_s}{R^2}$$

3-18. Assuming that a silicon photovoltaic detector is biased as shown
below:

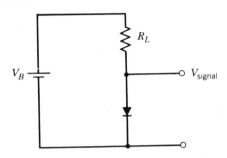

The *V–I* characteristic of the detector is shown below.

a. Find the operating points, the blackbody current responsivity,
and blackbody voltage responsivity if $V_B = -7$ V and $R_L = 35$ kΩ.
b. Repeat (a) for $V_B = -7$ V and $R_L = 450$ kΩ.
c. What can be inferred about the value of R_L to maximize
current responsivity?

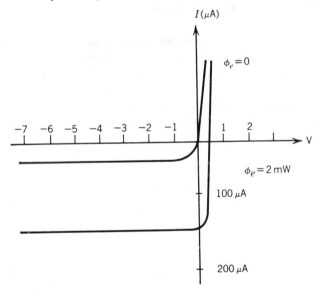

CHAPTER 4

PHOTOCONDUCTORS

4-1 INTRODUCTION

A photoconductor exhibits a change in conductance (resistance) when radiant energy (photons) is incident upon it. The radiant energy increases the conductance by producing more carriers in the detector. A photoconductor is operated in a mode in which an applied electric field produces a current that is modulated by additional carriers produced by photon excitation, that is, radiation quanta are absorbed and free (photogenerated) charge carriers are generated in the semiconductor. These additional carriers cause an increase in the conductivity of the semiconductor. This phenomenon is called photoconduction.

Photoconductors are made from semiconductor materials. The general characteristics of semiconductor photoconductors that make them different from thermal detectors are:

1. Time constant.
2. Spectral selectivity.
3. High sensitivity.

Photodetection time responses lie between those of fast photomultiplier tubes (10 nsec) and thermal detectors (50 msec), and typically are in the microsecond range under normal room-temperature background environments. The spectral responsivity is determined by the energy gap. Only photons that have energies greater than the energy gap will be absorbed and cause current to flow.

In this chapter, we will discuss the basic semiconductor/radiation interaction for the photoconductor. Then the basic equations describing the photoconductive detection process will be presented. The noise mechanisms that are present in photoconductors will be discussed and finally figures of merit such as NEP and D^* will be developed. The figures of merit will be discussed for two specific operation conditions, one being photon-noise-limited and the second being Johnson-noise-limited operation.

4-2   THEORY OF PHOTOCONDUCTOR DETECTION

4-2-1   *Background*

The mechanism of semiconductor photodetection can be explained by a discussion of the energy-band theory of semiconductors and the interaction of the radiation with the carriers in the detector. In solid-state theory the energy levels of the outer electrons of an atom are spread into nearly continuous bands by interaction with the neighboring atoms in the solid. Intrinsic semiconductors are characterized by a lattice in which all of the outer (valence) electrons are bound in covalent bonds. If ejected from a covalent bond, an electron will be in an essentially neutral environment and can thus become mobile and conduct current; the vacancy (hole) left by the electron also has mobility. If an electric field is applied, any free electrons will move in one direction and the corresponding holes in the valence band will move in the opposite direction.

The electronic energy states are represented in Fig. 4-1; the intrinsic semiconductor has a filled valence band and an empty conduction band, separated by a forbidden energy gap equal to the energy required to remove an electron from its covalent bond.

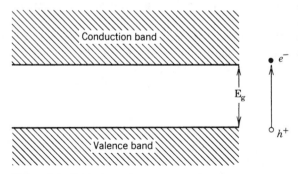

Figure 4-1   Intrinsic semiconductor energy-band diagram.

The energy required to cause an electron transition across the energy gap (E_g) can be supplied by thermal means ($kT > E_g$) or by radiation ($h\nu > E_g$), where $k =$ Boltzmann's constant, $h =$ Planck's constant, $T =$ temperature of semiconductor, and $\nu =$ optical radiation frequency. The radiation induced transitions form the basis for photoconductivity.

Table 4-1 lists some examples of intrinsic semiconductors that are photoconductors with their associated energy gaps.

Table 4-1

	E_g (eV)
PbSe	0.23
PbS	0.42
Ge	0.67
Si	1.12
CdSe	1.8
CdS	2.4

To minimize noise and thus maximize the sensitivity of the semiconductor as a radiation detector, the number of free carriers produced by thermal excitation must be made as small as possible. This can be done by cooling the material to a temperature T such that $kT \ll E_g$. However, there is a temperature range above absolute zero in which the semiconductor properties are essentially unchanged by further cooling. .

Free carriers are produced only when the radiation photons have sufficient energy to cause the electrons to cross the energy gap. Therefore, there is a limit on the wavelength response to which a given semiconductor will detect radiation.

The minimum optical frequency, ν, photon that will produce a free electron from the covalent bond is $\nu = E_g / h$. If one rewrites this limiting condition in terms of wavelength,

$$\lambda_c = \frac{1.24}{E_g} \tag{4.1}$$

where $\lambda_c =$ maximum wavelength of radiation to produce an electronic transition,

$\quad E_g =$ energy gap in eV.

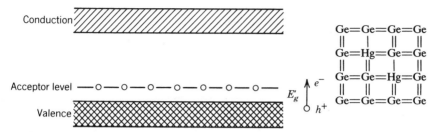

Figure 4-2   Extrinsic P-type semiconductor energy diagram.

As seen from Eq. (4.1) detectors used at longer wavelengths require smaller energy gaps (E_g). The wavelength response of a semiconductor material can be controlled by the doping of the intrinsic semiconductor (Levinstein, 1959). Therefore, by deliberate introduction of impurities into the semiconductor material the forbidden energy gap can be altered. The material then becomes an extrinsic semiconductor, either a P type for the positive majority carrier (holes) or an N type if the majority carrier is negative (electrons). Unlike the intrinsic semiconductor, in which both electrons and holes produce a change in conduction, in extrinsic semiconductors only the majority carriers produce a change in conduction.

For example, a P-type semiconductor can be produced by doping germanium with a Group-III atom such as mercury (Ge : Hg). The mercury has three valence electrons forming covalent bonds with germanium. The fourth electron must be provided by the germanium. This produces an excess number of holes in the bulk material. The energy-band structure for such a P-type semiconductor is shown in Fig. 4-2.

The result of the doping (addition of impurity atoms) is that less energy is required for an electron to jump from the valence band to the acceptor level. Therefore, the long-wavelength cutoff has been extended to a larger value. The electrons are trapped in the acceptor level and the holes (majority carriers) are mobile and cause conduction changes.

An N-type intrinsic semiconductor is produced by introducing impurity atoms of group V such as arsenic in silicon. The fifth valence electron of arsenic is bound to its nucleus, but only weakly, because of the shielding by other electrons in the silicon lattice. These electrons reside in a donor level within the energy gap as shown in Fig. 4-3, and can be excited into the conduction band by the absorption of energy.

In the N-type semiconductor a photon will cause an electron to jump from the donor level to the conduction band, thus causing conduction by electrons (majority carriers). A corresponding hole is trapped in the donor level.

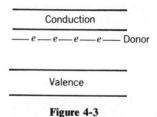

Figure 4-3

Table 4-2 lists some extrinsic semiconductors used as photoconductors. Magnesium doped silicon (Si:Mg) is an interesting material in that the spectral response extends further than expected (Lin, 1982).

Table 4-2

Examples of Extrinsic Photoconductive Detectors

	E_g (eV)	Conductor Type
Ge:Hg	0.09	P
Ge:Cu	0.041	P
Ge:Cd	0.06	P
Si:As	0.0537	N
Si:Bi	0.0706	P
Si:P	0.045	N
Si:In	0.165	P
Si:Mg	0.087	P

Since the energy gap for the extrinsic semiconductor is smaller, the effects of thermal carrier generation are more severe. Therefore, the operating temperature required to eliminate thermally generated carriers must be lower.

In order to keep the thermally ionized carrier conduction down, the temperature must be lower than T_c, where T_c is defined as (Beynon and Lamb, 1980; Blakemore, 1974):

$$T_c = \frac{E_g}{k \ln[v\sigma_c N_V / \sigma_{ph} E_p^B]} \tag{4.2}$$

where v = thermal drift velocity,

  σ_c = free electron capture cross section,

  σ_{ph} = photoionization cross section,

  N_V = acceptor/donor concentration,

  E_p^B = photon irradiance.

For the case of large photon irradiance values the expression reduces to the more straightforward expression of

$$T_c \leq \frac{E_g}{k} \qquad (4.3)$$

where the less than equal sign is introduced to say a lower temperature can be used.

Recalling Eq. (4.1) for the relationship between energy gap (E_g) and wavelength cutoff (λ_c) one can see that Eq. (4.3) shows that $\lambda_c T_c$ is less than or equal to a constant. Therefore, as the wavelength response gets longer, the cooling requirements become more severe. A rule of thumb is that a 10-μm cutoff requires liquid nitrogen temperature (77 K) (Beyen and Pagel, 1966).

4-2-2   *Theory of Photoconductive Detection*

In order to put the previous discussion of photoconductive detection on a firmer mathematical basis, the response to a radiation signal will be derived. Mathematically one can express the conductivity (ohm^{-1} cm^{-1}) of a photoelectric semiconductor device which is not illuminated as:

$$\sigma_0 = N_0 \mu_e q + P_0 \mu_h q \qquad (4.4)$$

where N_0 and P_0 = free carrier concentrations of electron and holes, respectively (# cm^{-3}),

  μ_e and μ_h = mobilities of electron and holes (cm^2 sec^{-1} V^{-1}),

  q = charge on electron, 1.6×10^{-19} C.

If ΔN and ΔP are the concentration of excess electrons and hole carriers generated by the absorption of photons, the conductivity expression will be

$$\sigma = q[\mu_e(N_0 + \Delta N) + \mu_h(P_0 + \Delta P)] \qquad (4.5)$$

The effect of this excess carrier concentration is to raise the conductivity thus producing a viable detection mechanism for absorbed radiation. Conceptually, there is a constant conductivity term and a varying conductivity term,

$$\sigma = \sigma_0 + \Delta\sigma \qquad (4.6)$$

where σ_0 is as shown in Eq. (4.4) and

$$\Delta\sigma = q(\mu_e\Delta N + \mu_h\Delta P) \tag{4.7}$$

Since ΔN and ΔP are the change in carrier concentrations due to absorbed photons, the number of electrons is equal to the number of holes, $\Delta P = \Delta N$. So with no loss of generality we can rewrite Eq. (4.7) as

$$\Delta\sigma = q(\mu_e + \mu_h)\Delta N \tag{4.8}$$

where $q =$ charge of an electron $(1.6\times10^{-19}\text{ C})$,
  $\mu_e =$ electron mobility $(\text{cm}^2\text{ sec}^{-1}\text{ V}^{-1})$,
  $\Delta N =$ change in carrier concentration due to absorbed radiation.

The change in carrier concentration, ΔN, is caused by signal photons being absorbed and can be expressed as

$$\Delta N = \frac{\eta\Delta\phi_P^S\tau_L}{A_d W} \tag{4.9}$$

where $\eta =$ quantum efficiency,
  $\Delta\phi_P^S =$ signal photons (photon sec^{-1}),
  $\tau_L =$ carrier lifetime,
  $A_d =$ detector active area (see Fig. 4-4),
  $W =$ detector thickness (see Fig. 4-4).

This assumes that a change in the number of carriers per unit volume is due to signal photons producing carriers with a lifetime (τ_L) in a detector volume of ltW as shown in Fig. 4-4. The quantum efficiency in Eq. (4.9) is

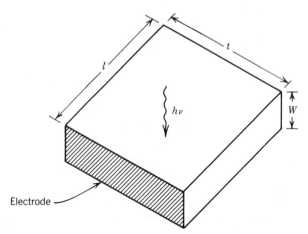

Figure 4-4 Detector geometry. Signal flux incident on surface lt.

the probability of independent photocarrier generation in the detector per incident photon within the spectral sensitivity of the device. An assumption has been made that the thickness (t) is sufficiently large to capture all the photons. Now, by substitution, we can get the relative change in conductivity by

$$\frac{d\sigma}{\sigma} = \frac{q(\mu_e + \mu_h)}{\sigma} \frac{\eta \Delta \phi_P^S \tau_L}{A_d W} \tag{4.10}$$

This is the change in conductivity due to the presence of signal photons ($\Delta \phi_P^S$) on the detector's active area.

It is more conventional to use resistance from an engineering viewpoint. Therefore, the resistance of the detector geometry shown in Fig. 4-4 can be expressed as

$$R_d = \frac{l}{\sigma W t} \tag{4.11}$$

where l = length of photoconductor (distance between electrodes),
$\quad \sigma$ = conductance [(ohm cm)$^{-1}$],
$\quad W$ = width of photoconductor (photon absorption path),
$\quad t$ = thickness of photoconductor.

The differential form of Eq. (4.11) is

$$dR_d = -\frac{l \, d\sigma}{\sigma^2 W t} = -R_d \frac{d\sigma}{\sigma} \tag{4.12}$$

This can be thought of as the relative change in resistance having an opposite slope than that of the relative change in conductance, therefore the negative sign in the expression.

By direct substitution into Eq. (4.10), the change in resistance due to incident photon flux is

$$dR_d = -R_d \frac{q(\mu_e + \mu_h)}{\sigma} \frac{\eta \Delta \phi_P^S \tau_L}{A_d W} \tag{4.13}$$

The photon signal ($\Delta \phi_P^S$) can be expressed in watts of radiant power assuming that the signal is monochromatic:

$$\Delta \phi_P^S = \frac{\Delta \phi_e^S}{hc / \lambda} \tag{4.14}$$

where $\Delta\phi_e^S$ = signal radiant power (W),
  h = Planck's constant,
  c = speed of light,
  λ = wavelength of radiation.

The change in resistance due to radiation can be expressed as shown in Eq. (4.13).

In order to sense this photogenerated resistance change the device must be biased. A standard interface circuit consists of a DC bias and a load resistor as shown in Fig. 4-5.

The output DC voltage (V_0) shown in Fig. 4-5 is

$$V_0 = \frac{V_B R_L}{R_d + R_L} \tag{4.15}$$

The change in output voltage due to a change in detector resistance (R_d) is found by differentiating Eq. (4.15):

$$dV_0 = \frac{-V_B R_L \, dR_d}{(R_d + R_L)^2} \tag{4.16}$$

where dV_0 is the AC voltage signal, caused by signal photons, superimposed on the DC level V_0. Substituting from Eqs. (4.13) and (4.14) into (4.16) we obtain

$$dV_0 = \frac{V_B R_L R_d}{(R_L + R_d)^2} \cdot \frac{q\lambda\eta\tau_L(\mu_e + \mu_h)}{\sigma h c A_d W} \Delta\phi_e^S \tag{4.17}$$

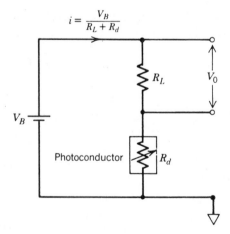

Figure 4-5 Standard photoconductor biasing configuration.

The signal voltage can be rewritten using the DC current flowing through the circuit shown in Fig. 4-5. In addition, the voltage responsivity ($\mathscr{R}_V$) can now be written directly from Eq. (4.17):

$$\mathscr{R}_V = \frac{dV_0}{\Delta\phi_e^s} = \frac{iq\lambda\eta\tau_L(\mu_e + \mu_h)}{\sigma hcA_dW}\frac{R_LR_d}{R_L + R_d} \tag{4.18}$$

This is the spectral responsivity for a photoconductor.

As seen by Eq. (4.18), the voltage responsivity is a function of the load resistor/detector resistance in parallel. The current responsivity expression can be derived using the relationship between di and dV. The expression for current responsivity becomes

$$\mathscr{R}_i = \frac{iq\lambda\eta\tau_L(\mu_e + \mu_h)}{\sigma hcA_dW} \cdot \frac{R_d}{R_L + R_d} \tag{4.19}$$

where i = DC current through detector,
q = electron charge,
λ = wavelength,
τ_L = carrier lifetime,
η = quantum efficiency,
μ_e, μ_h = mobilities,
σ = conductivity,
h = Planck's constant,
c = speed of light,
A_d = active area of detector,
W = thickness.

Both the voltage responsivity and the current responsivity expressions, Eqs. (4.18) and (4.19), respectively, assume that both electron and hole carriers are present in an intrinsic photoconductor. For an extrinsic semiconductor, only one carrier is present, and the above expressions are modified by dropping the mobility term that is inappropriate. It must be remembered that this expression is only valid for low-frequency operation, that is, where the dielectric relaxation or recombination lifetime effects are not of concern.

The parameters that the detector manufacturer needs to increase to get a higher responsivity are as follows:

1. Quantum efficiency by creating a greater photoionization cross section.
2. Carrier lifetime.

3. Mobility (recall N-type carriers have higher mobility than P-type carriers).

4. Current (by increasing the bias voltage levels across the detector).

In addition, the ideal responsivity is a linear function of wavelength. However, for actual devices if one plots the responsivity versus wavelength, it will be nonlinear at the short- and long-wavelength operating limits (Fig. 4-6).

There are several reasons for the departure from ideal responsivity. In the short-wavelength region, the Fresnel reflection loss $\{[(n-1)/(n+1)]^2\}$, due to the index of refraction dependence on wavelength, causes some loss of incident photons. Also, the absorption of short-wavelength photons occurs nearer the surface where there is a higher density of surface-state traps (Willardson and Beer, 1977). The long-wavelength cutoff of responsivity is primarily due to the wavelength dependent transmission of the material of the following form:

$$\tau = [1 - \rho(\lambda)]e^{-\alpha(\lambda)t} \tag{4.20}$$

where $\rho(\lambda)$ = reflection coefficient,
$\alpha(\lambda)$ = absorption coefficient,
t = thickness.

The absorption coefficient and reflection coefficient are functions of wavelength that, when used with the thickness, predict the rolloff responsivity for long wavelength (Finkman and Nemirovski, 1979).

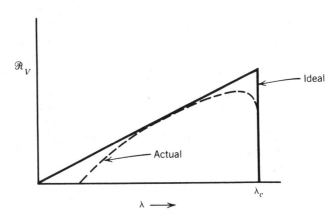

Figure 4-6   Responsivity versus wavelength.

4-2-3   *Photoconductive Gain*

Let us regroup the terms in the current responsivity expression, Eq. (4.19), to look like the expression derived for the photovoltaic detectors in Section 3-3:

$$\mathscr{R}_i = \frac{q\lambda\eta}{hc}\, G \tag{4.21}$$

where by direct comparison to Eq. (4.19)

$$G = \frac{i\tau_L(\mu_e + \mu_h)}{\sigma A_d W} \tag{4.22}$$

This term is called the photoconductive gain (G). To simplify the expression, we will consider only extrinsic photoconductors, so that μ will replace $\mu_e + \mu_h$ in Eq. (4.21). The photoconductive gain can be rewritten as*

$$G = \frac{\tau_L \mu E}{l} \tag{4.23}$$

where τ_L = free carrier lifetime,
   μ = mobility of carrier present,
   l = interelectrode spacing,
   E = electric field strength in detector (V cm^{-1}).

This is the expression most commonly presented for photoconductive gain (Rose, 1963).

The carrier transit time (τ_t) across the photoconductor for a carrier can be expressed as

$$\tau_t = \frac{l}{\mu E} \quad \left[\frac{\text{distance}}{\text{velocity}}\right] \tag{4.24}$$

The magnitude of the photocurrent is determined by the drift velocity across the semiconductor (interelectrode spacing) and the carrier lifetime. Conceptually, one can think about the photogenerated carrier having a lifetime long enough to traverse the semiconductor electrodes. If the carrier lifetime is long, the carrier may traverse the detector several times before it recombines. Thus the photoconductive gain can be expressed as the ratio of carrier lifetime (τ_L) to detector transit time (τ_t)

*Note: The electric field is the voltage across the detector divided by the interelectrode spacing (l in Fig. 4-4). From Eq. (4.11) for resistance, Eq. (4.23) can be rewritten in terms of voltage across the detector.

$$G = \frac{\tau_L}{\tau_t} \tag{4.25}$$

For a linear extrinsic photoconductor with uniform resistance and uniform applied electric field, the current flow is due to majority carriers which recombine at the stationary electron–hole pair. In this case, the photoconductive gain is proportional to the applied electric field and is typically about 0.5–1.0. For low background levels ($E_P \approx 0$), the photoconductive gain approaches the value of 0.5. The explanation of this is that as an AC photogenerated electron leaves the detector it leaves behind a space-charge buildup which impedes another electron from entering the detector from the opposite electrode. This space-charge buildup for the AC case gives rise to a dielectric relaxation time constant (τ_ρ), which is given by

$$\tau_\rho = \rho\varepsilon \tag{4.26}$$

where ρ = detector resistivity,
ε = dielectric constant.

For low-background irradiance levels the resistivity becomes very large and the dielectric relaxation time is very long (i.e., seconds). Since typical optical modulation frequencies are above τ_ρ^{-1}, the electrical signal from the detector is a low-pass filtered version of the modulation waveform. It has been shown that if the detector contacts obey a certain model with solvable boundary conditions, the photoconductive gain is exactly 0.5 under low-background conditions (Perry, 1981).

The intrinsic PbS detector has a photoconductive gain much greater than unity. This is caused by natural or induced discontinuities, which produce localized regions of high electric field. These high electric fields cause an increase in carrier velocity. This allows the production of a secondary current that is larger than the photogenerated current. This secondary current enters from the electrodes, and it is a function of the acceleration of the photoelectrons, the space charge of traps, and the natural and induced discontinuities that exist within the detector material. Therefore, a single photon can produce more than a single electron–hole pair. The photoconductive gain can also be thought of as the number of carriers passing through the detector per absorbed photon. Therefore, it is probably more correct to use the photoconductive gain–quantum efficiency ($G\eta$) product.

4-2-4 *Responsivity versus Temporal Frequency*

The expressions of responsivity [Eqs. (4.18) and (4.19)] were for low temporal frequency operation. If one considers a pulse of radiation turning on at time zero, then the continuity equation implies the following differential equation:

$$\frac{d\Delta P}{dt} = g - \frac{\Delta P}{\tau_L} \qquad (4.27)$$

where ΔP = number of excess holes,

τ_L = carrier (hole) lifetime,

g = generation rate of excess carriers by photon absorption.

Solving Eq. (4.27):

$$\Delta P = \tau_L g [1 - e^{-t/\tau_L}] \qquad (4.28)$$

for the boundary condition that $\Delta P = 0$ at time $t = 0$. The detector exhibits exponential rise and fall in response to instantaneous changes in the irradiance.

The frequency response can be found from the Fourier transform of an instantaneous time pulse (Gaskill, 1978, p. 202)

$$\Delta P(f) = 2\tau_L g \frac{1}{1 + j2\pi f \tau_L} \qquad (4.29)$$

$$= \frac{2\tau_L g}{\sqrt{1 + (2\pi f \tau_L)^2}} \qquad (4.30)$$

The above expression exhibits the classical 6 dB per octave rolloff of a single-pole filter, where the charge carrier lifetime determines the cutoff frequency. This expression for responsivity as a function of frequency is shown graphically in Fig. 4-7.

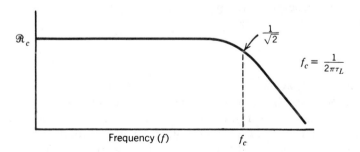

Figure 4-7 Frequency response.

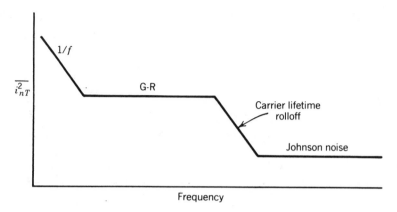

Figure 4-8   Typical photoconductor mean square noise spectrum.

4-3 NOISES

The noise level of the detector determines the limit to sensitivity, and it must be considered in order to determine the detectable incident power.

The noises that are present in photoconductive applications are $1/f$ noise, Johnson noise, and generation–recombination (photon) noise. Figure 4-8 shows a classical noise spectrum for a photoconductive detector, the $1/f$ noise is dominant at low frequencies, generation–recombination (G-R) noise dominates at midband, and Johnson noise dominates at high frequencies. The transition frequencies from one type of noise to another vary with the detector material used. Additionally, the $1/f$ to G-R transition varies between detectors made of the same material. For example, it can vary from 5 to 200 Hz for doped silicon and HgCdTe detectors depending on contact technology and techniques used in fabrication. The carrier lifetime break point is typically near 1 MHz, which again varies for various detector types.

4-3-1 $1/f$ Noise

The physical mechanism that produces this noise source is not well understood. The $1/f$ dependence (higher noise level at lower frequencies) holds for the noise power, the noise voltage varying as one over the square root of frequency ($1/\sqrt{f}$). Since photoconductors require a bias current (see Fig. 4-5), there will always be $1/f$ noise present. The following equation has been empirically obtained for the mean square noise current

$$\overline{i_{1/f}^2} = \frac{B_1 I_b^2 \Delta f}{f^\beta} \qquad (4.31)$$

where B_1 = proportionality constant,
$\quad I_b$ = DC current through detector,
$\quad \Delta f$ = electrical bandwidth,
$\quad f$ = frequency,
$\quad \beta$ = constant, which is usually 1.

For a photoconductor at low frequencies the dominant noise exhibits the $1/f$ dependence of Eq. (4.31). As the frequency of interest is increased, this component drops below the G-R or Johnson noise for low photon flux and cryogenically cooled detectors.

A good theoretical formation of $1/f$-type noise has not been developed, although it has been studied extensively for three decades (MacFarlane, 1950, Richardson, 1950). It has been found that the noise is a strong function of the quality of electrical contact with the semiconductor. Good ohmic contacts produce less $1/f$ noise. It has also been shown that the level of $1/f$ noise is directly proportional to the density of surface states for a metal–insulator–semiconductor contact (Deal et al., 1967).

The $1/f$ noise component does not present a fundamental limit to sensitivity. Careful surface preparation and electrical contacting methods reduce this noise to negligible levels. In any case, the development of a low $1/f$ noise photoconductor *remains an art* rather than a science. In this chapter we will consider that the $1/f$ noise is negligible compared to other, more fundamental, noises in photoconductors.

4-3-2 *Johnson Noise*

Johnson noise is caused by the thermal agitation of electrons in a resistor. It is also called Nyquist or thermal noise, but is most often called Johnson noise after the worker who first observed it (Johnson, 1928). It occurs in all resistive materials.

For a photoconductor of resistance R_d, the expression for Johnson noise mean square current is

$$\overline{i_j^2} = \frac{4kT\Delta f}{R_d} \qquad (4.32)$$

where k = Boltzmann constant,
$\quad T$ = temperature,
$\quad R_d$ = detector resistance,
$\quad \Delta f$ = electrical bandwidth.

For the typical biasing configuration shown in Fig. 4-5 the total Johnson mean square current is

$$\overline{i_{JT}^2} = 4k\Delta f\left(\frac{T_d}{R_d} + \frac{T_L}{R_L}\right) \tag{4.33}$$

where T_d, T_L = temperatures of detector and load,
$\quad R_d$, R_L = resistances of detector and load.

4-3-3   Generation–Recombination Noise

The generation–recombination (G-R) noise is caused by the fluctuations in generation rates, recombination rates, or trapping rates in a photoconductor (semiconductor) thus causing fluctuations in free carrier current concentration. The fluctuation in rate of generation and recombination is affected by two processes, thermal excitation of carriers and photon excitation (Van Vliet, 1967). The G-R noise has been studied in detail for infrared detectors by Smith (1982), and the expression for the combination of G-R noise due to both photon and thermal excitation is given by

$$\overline{i_{G-R}^2} = 4q(q\eta E_p^B A_d G^2 + qg_{th}G^2)\,\Delta f \tag{4.34}$$

where q = electron charge,
$\quad \eta$ = quantum efficiency,
$\quad E_p^B$ = photon irradiance,
$\quad A_d$ = detector area,
$\quad G$ = photoconductive gain,
$\quad g_{th}$ = thermal generation rate.

The second term in Eq. (4.34) is the fluctuation in rate due to the thermal generation of carriers in the photoconductor. If the device is cooled sufficiently, thermal generation will decrease, so that it can be neglected. Therefore, when discussing G-R noise we will be talking about the photon noise that dominates G-R noise for cooled detectors. The expression for photon noise (G-R noise) that will be used for a photoconductor is

$$\overline{i_{G-R}^2} = 4q^2 G^2 \eta E_p A_d\, \Delta f \tag{4.35}$$

4-3-4   Summation of Noise Sources

Figure 4-8 showed the dominant noise in various frequency regions. The noises add in quadrature (variances add). Therefore, for any frequency region in Fig. 4-8, the noise was the summation of all the noise as shown below:

$$\overline{i_{nt}^2} = \overline{i_{i/f}^2} + \overline{i_J^2} + \overline{i_{G-R}^2}$$

and one of the terms was the dominant noise:

$$\overline{i_{nt}^2} = \left[\frac{B\,I^2\Delta f}{f} + \frac{4kT_D\Delta f}{R_d} + \frac{4kT_L\Delta f}{R_L} + 4q^2G^2\eta E_p A_d \Delta f\right] \quad (4.36)$$

4-4 FIGURES OF MERIT FOR PHOTOCONDUCTORS

4-4-1 *Photon Noise Limited*

Ideally, the noise related to the photon irradiance on the detector determines the noise. This is the best possible situation. In order to have that situation, the G-R noise (photon noise) must be the dominant noise, therefore,

$$\overline{i_{\text{G-R}}^2} \gg \overline{i_J^2}$$

or from Eqs. (4.33) and (4.35)

$$4q^2G^2(\eta E_p A_d \Delta f) \gg 4k\Delta f\left(\frac{T_d}{R_d} + \frac{T_L}{R_L}\right) \quad (4.37)$$

The $1/f$ noise is neglected, since it depends on manufacturing techniques and operating conditions and is not a fundamental noise limit. If the load resistor is mounted on the detector heat sink, the temperatures are equal, $T_d = T_L = T$. This situation also helps reduce the Johnson noise, because the photoconductor is usually cryogenically cooled.

Solving Eq. (4.37) for the temperature/resistance ratio:

$$\frac{T}{R_{\text{eff}}} \ll \frac{q^2 G^2 (\eta E_p A_d)}{k} \quad (4.38)$$

where T = temperature of detector/load resistor,
R_{eff} = parallel equivalent resistance of R_d and R_L $[R_L R_d/(R_L + R_d)]$.

If the detector temperature divided by the equivalent resistance is less than a constant which is fixed by parameters of the system, then the detector will achieve photon-noise-limited operation. In order to be photon noise limited the ratio of temperature to resistance must go down as the photon irradiance (E_p) decreases. This sets some limits on the load resistor and its temperature. Typically, for high-performance extrinsic IR detectors, the load resistance (R_L) is much smaller than the detector resistance so that the equivalent resistance is actually the load resistor value, $R_{\text{eff}} \cong R_L$; however, for intrinsic photoconductors such as HgCdTe, $R_{\text{eff}} = R_d$.

If the photon irradiance incident on a photon-noise-limited detector is due primarily to background photons, then the detector is said to be a background-noise-limited infrared photodetector (BLIP). For this BLIP operation the spectral NEP can now be expressed as

$$\mathrm{NEP}(\lambda, f) = \frac{\sqrt{\overline{i^2_{G\text{-}R}}}}{\mathcal{R}_i} \qquad (4.39)$$

where $\sqrt{\overline{i^2_{G\text{-}R}}}$ = photon current noise,
$\mathcal{R}_i$ = current responsivity.

Or, substituting,

$$\mathrm{NEP}(\lambda, f) = \frac{2qG[\eta E_p^B A_d \Delta f]^{1/2}}{(q\lambda\eta/hc)G}$$

$$= \frac{2hc}{\lambda}\left[\frac{E_p^B A_d \Delta f}{\eta}\right]^{1/2} \qquad (4.40)$$

where h = Planck's constant,
c = speed of light,
λ = wavelength,
E_p^B = background photon irradiance,
A_d = detector area,
Δf = electrical bandwidth,
η = quantum efficiency.

This is the BLIP spectral NEP for a photoconductor. The corresponding BLIP spectral D^* (see Section 2-4) is

$$D^*(\lambda, f) = \frac{\sqrt{A_d \Delta f}}{\mathrm{NEP}} \qquad (4.41)$$

or by substitution from Eq. (4.40)

$$D^*(\lambda, f) = \frac{\lambda}{2hc}\sqrt{\frac{\eta}{E_p^B}} \qquad (4.42)$$

where λ = wavelength,
h = Planck's constant,
c = speed of light,
η = quantum efficiency,
E_p^B = background irradiance.

For this BLIP operation of a photoconductor the highest peak spectral D^* is plotted in Fig. 4-9. Also included in Fig. 4-9 are some common

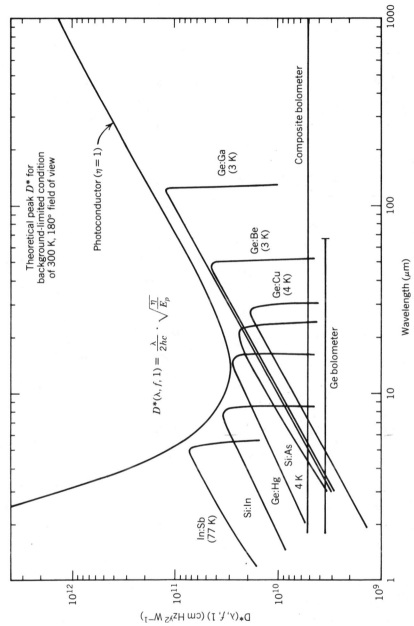

Figure 4-9 Typical D^* for selected photoconductor.

$$D^*(\lambda, f, 1) = \frac{\lambda}{2hc} \cdot \sqrt{\frac{\eta}{E_p}}$$

Theoretical peak D^* for background-limited condition of 300 K, 180° field of view

Photoconductor ($\eta = 1$)

Ge:Ga (3 K)

Ge:Be (3 K)

Ge:Cu (4 K)

Si:In

Ge:Hg

Si:As 4 K

In:Sb (77 K)

Ge bolometer

Composite bolometer

Wavelength (μm)

$D^*(\lambda, f, 1)$ (cm Hz^{y_2} W^{-1})

extrinsic photoconductors and their corresponding operating temperatures.

4-4-2 *Johnson Noise Limited*

From Eq. (4.42), the BLIP spectral D^* gets larger (better) as the background irradiance goes down. If one plots the D^* versus background irradiance as shown in Fig. 4-10, a limit is reached as the background photon irradiance is decreased.

If the background photon irradiance is reduced to zero, no further improvement in D^* can be realized since the condition of Eq. (4.38) is not satisfied. In fact, the Johnson noise becomes the dominant noise. The corresponding spectral NEP is

$$\text{NEP}(\lambda, f) = \frac{\sqrt{\overline{i_j^2}}}{\mathscr{R}_i}$$

or

$$\text{NEP}(\lambda, f) = \frac{\sqrt{4k\Delta f[T_L/R_L + T_d/R_d]}}{(q\lambda\eta/hc)G} \tag{4.43}$$

If we make similar assumptions to those in Section 4-4-1, namely, $T_L = T_d = T$ and the load resistor is much less than the detector resis-

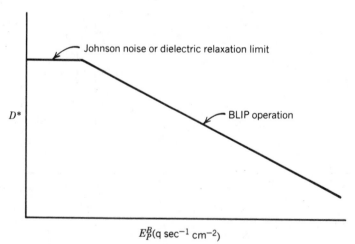

$$E_p^B(\text{q sec}^{-1}\text{ cm}^{-2})$$

Figure 4-10 D^* versus photon background irradiance.

tance, we can rewrite Eq. (4.43)

$$\text{NEP}(\lambda, f) = \frac{hc}{q\lambda\eta G}\left(\frac{4k\Delta fT}{R_{\text{eff}}}\right)^{1/2} \tag{4.44}$$

where h = Planck's constant,
c = velocity of light,
k = Boltzmann's constant,
Δf = noise bandwidth,
$T = T_L = T_d$ = temperatures,
R_{eff} = effective resistance,
q = electronic charge,
λ = wavelength,
η = quantum efficiency,
G = photoconductive gain.

The corresponding spectral D^* for the Johnson-noise-limited case is

$$D^* = \frac{\sqrt{A_d \Delta f}}{\text{NEP}}$$

$$D^* = \frac{q\lambda\eta}{2hc}\left(\frac{G^2 R_{\text{eff}} A_d}{kT}\right)^{1/2} \tag{4.45}$$

where q = electronic charge,
λ = wavelength,
η = quantum efficiency,
h = Planck's constant,
c = speed of light,
G = photoconductive gain,
R_{eff} = parallel equivalent resistance,
A_d = detector area,
k = Boltzmann's constant,
T = temperature of detector/load resistor.

It should be noted that the resistance–area product determines the plateau level (transition to Johnson-noise-limited performance) in Fig. 4.10. That is why a high RA product is required for high-performance detectors. It should also be noted that the D^* is linearly related to photoconductive gain, and that is how HgCdTe can have a high D^* value with a low RA product.

When operating in the Johnson noise limit, the theoretical achievable D^* is determined by the load resistor value ($R_d \gg R_L$) and is in fact the

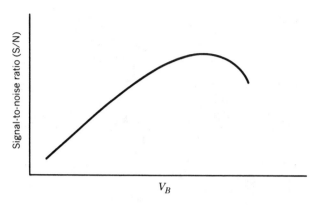

Figure 4-11 Signal-to-noise ratio versus detector bias voltage.

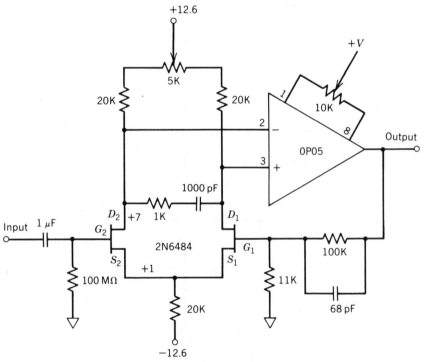

Figure 4-12 Low noise voltage preamplifier.

108

same expression as that of the photovoltaic detector where the effective resistance is the junction resistance. The difference lies in the photoconductive gain. For the low-background case, in order for the photoconductor ($G = 0.5$) to have equivalent performance to the photovoltaic detector, requires that the load resistor be four times the junction resistance of the photovoltaic detector.

In practice the detector must be optimally biased and used with a low-noise preamplifier. If a plot of signal-to-noise ratio is made for various detector bias values, a plot as shown in Fig. 4-11 is obtained. The S/N drops off after going through a peak value due to the increase in noise above that obtained at optimum bias. This increase in noise is caused by the breakdown of the semiconductor or the electrodes.

In order to ensure detector noise limited operation, a low-noise preamplifier must be used. Figure 4-12 shows a voltage-gain amplifier, which has been used successfully. For this type of preamplifier the selection of an input JFET with a low-noise voltage is important. A current-mode preamplifier has also been used successfully with photoconductive detectors. For further description the reader is referred to Dereniak, Joyce, and Capps (1977).

REFERENCES

Beyers, W., et al., "Cooled Photoconductive Infrared Detectors," *J. Opt. Soc. Am.* **49**, 686 (1959).

Beynon, J. D. E. and D. R. Lamb, *Charge Coupled Devices and Their Applications*, McGraw-Hill, United Kingdom, 1980.

Beyen, W. J. and B. R. Pagel, "Cooling Requirements for Intrinsic Photoconductive Infrared Detectors," *IR Phys.* **6**, 161 (1966).

Blakemore, J. S., *Solid State Physics*, 2nd ed., Saunders, Philadelphia, 1974.

Bube, R. H., *Photoconductivity of Solids*, Wiley, New York, 1960.

Deal, B. E., M. Sklar, A. S. Grove, and E. H. Snow, "Characteristics of the Surface State Charge of Thermally Oxidized Silicon," *J. Electrochem. Soc.* 114 (1967).

Dereniak, E. L., R. R. Joyce, and R. W. Capps, "Low Noise Preamplifier for Photoconductive Detectors," *Rev. Sci. Instrum.* **48**(4) (April, 1977).

Gaskill, J. D., *Linear Systems, Fourier Transforms and Optics*, Wiley-Interscience, New York, 1978.

Finkman, E. and Y. Nemirovski, "Infrared Optical Absorption of HgCdTe," *J. Appl. Phys.* **50**(6) (1979).

Hudson, R. D., *Infrared System Engineering*, Wiley, New York, 1969.

Johnson, J. B., "Thermal Agitation of Electricity in Conductors," *Phys. Rev.* **32**, 97–109 (1928).

Kruse, P. W., L. D. McGlauchlin, and R. B. McQuistan, *Elements of Infrared Technology*, Wiley, New York, 1963.

Levinstein, H., "Extrinsic Detectors," *Appl. Opt.* **4**, 639 (1965).

Levinstein, H., "Impurity Photoconductivity in Germanium," *Proc. IRE* **47**, 1478 (1959).

Lin, A. L., "Electrical and Optical Properties of Magnesium Diffused Silicon," *J. Appl. Phys.* **53**(10), 6989 (1982).

MacFarlane, G. G., "A Theory of Contact Noise in Semiconductors," *Proc. Phys. Soc.* (*London*) **63B**, 807–814 (1950).

Nicollian, E. H. and H. Melchior, "A Quantitative Theory of 1/f Type Noise due to Interface States in Thermally Oxidized Silicon," *Bell Syst. Tech. J.* **46**, 1935 (1967).

Perry, C., Private Communication, Aerojet Electrosystems, Azusa, CA (1981).

Richardson, J. M., "The Linear Theory of Fluctuations Arising from Diffusional Mechanisms—An Attempt at a Theory of Contact Noise," *Bell Syst. Tech. J.* **29**, 117–141 (1950).

Rose, A., *Concepts in Photoconductivity and Allied Problems*, Interscience Publishers, New York, 1963.

Smith, D. L., "Theory of Generation–Recombination Noise in Intrinsic Photoconductors," *J. Appl. Phys.* **53**(10), 7051 (1982).

Van Vliet, K. M., "Noise Limitations in Solid State Photodetectors," *Appl. Opt.* **6**, 7 (1967).

Willardson, R. K., and A. C. Beer, *Semiconductors and Semimetals*, *Vol. 12*, *Infrared Detector II*, Academic Press, New York, 1977.

PROBLEMS

4-1. If the blackbody responsivity $\mathcal{R}(T, f)$ of a photoconductor is 5 A/W and its resistance is 10^6 Ω, what is the NEP and D^* (assume room-temperature operation, Johnson noise limited, active area = 1 mm × 1 mm, and $\Delta f = 10$ Hz).

4-2. Explain any correlation of cutoff wavelength and operation temperature for a photoconductor.

4-3. Why is the impedance (DC resistance) of HgCdTe not a function of background (E_p^B)?

4-4. Why does peak spectral D^* versus detector temperature for PbS and PbSe photoconductors have a maximum?

4-5. In choosing a preamplifier for a photoconductor, which noises are most important for a low-impedance detector? High-impedance detector?

4-6. How does the BLIP D^* vary with operating temperature for photoconductors?

4-7. a. What is the range of values for the temperature load resistor quotient (T_L/R_L) to have the detector/preamplifier remain BLIP for a Si:As photoconductor which has quantum efficiency = 0.5, background irradiance $(E_p^B) = 10^{10}$ q sec^{-1} cm^{-2}, photoconductive gain = 0.5, $\Delta f = 1$, detector area = 10^{-1} cm^2.

b.   What would the range of load resistor values have to be for T_L equal to 10 K?

4-8.   What is the expression of the BLIP $D^*(\lambda, f)$ for a Ge:Cu in terms of the detector resistance?

4-9.   Discuss the significance of the resistance area product (RA).

4-10.   Compare and contrast high-impedance photoconductors versus low-impedance photoconductors.

4-11.   Explain photoconductive gain in words. Derive the expression $G = \tau_L/\tau_t$ (carrier lifetime τ_L to carrier transit time τ_t). Show that it is also

$$G = \frac{i\tau_L\mu}{\sigma A_d W}$$

4-12.   Make a plot of $D^*(\lambda_p, f)$ versus background photon irradiance (E_p^B) (log–log) on a Hg$_{0.8}$Cd$_{0.2}$Te for backgrounds between 10^3 and 10^{18} q sec^{-1} cm^{-2} operating at 4 K. Discuss any deviations from the slope of $-1/2$ (i.e., any dominant noises).

4-13.   Calculate the Johnson noise voltage $(V_{n\bar{j}})$ across a parallel combination of resistors as shown $(\Delta f = 1000$ Hz$)$

a.

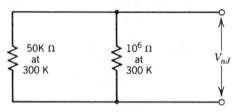

b.   Cool the 50 K to liquid–helium temperature.

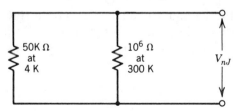

c.   Would cooling the 1 MΩ, resistor have a larger effect? Why?

4-14.   Why can the 1/f noise be eliminated in a photovoltaic detector but not in a photoconductor?

4-15. A photoconductive detector has a D^*_{BLIP} (10 μm, 1000) = 10^{11} cm $Hz^{1/2}$ W^{-1} with a 250 K background temperature and a 60° field of view (2θ). What is the quantum efficiency (η)?

4-16. Show how performance (D^*) can be improved by immersion of a detector into a hemispherical lens. Assume a Ge lens (n (index) = 4). Calculate the gain in D^* with the lens for:

   a. Internally noise limited detector.

   b. BLIP-limited condition.

4-17. A photoconductive detector which must operate at liquid-nitrogen temperature (77 K) is biased as shown using a room-temperature load resistor [assuming a constant signal current of I_s (10^9 A) and noise equivalent bandwidth of 1, $\Delta f = 1$]. What is the signal-to-noise ratio if the load resistor R_L is

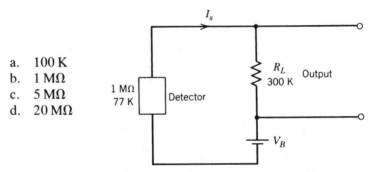

   a. 100 K
   b. 1 MΩ
   c. 5 MΩ
   d. 20 MΩ

   For this particular detector, what conclusion can be drawn about maximizing signal-to-noise ratio with regard to the load resistor requirements (value and temperature)?

4-18. Why does an extrinsic photoconductor require cooling to a lower temperature than an intrinsic photoconductor with the same cutoff wavelength?

4-19. What is the photoconductive gain for a HgCdTe photoconductor ($0.5 \times 0.5 \times 0.01$ mm^3) with a power dissipation of 1 mW and $\tau_L = 0.5$ μsec? $\mu_e = 3 \times 10^5$ [volt sec]$^{-1}$ cm^2, $N_o = 10^{14}$ cm^{-3}.

CHAPTER 5

PHOTOEMISSIVE DETECTOR THEORY

5-1 EXTERNAL PHOTOEFFECT

The external photoeffect involves ejecting electrons from the surface of a photocathode in contrast with the internal photoeffects discussed in Chapters 3 and 4. An applied electric field can be used in a vacuum device as shown in Fig. 5-1 to collect the photelectrons and create a photon-generated current in an external circuit. The current generated by the photon stream is the number of photons incident per second (ϕ_ρ) multiplied by the fraction of photons that produce electrons (η) multiplied by the electric charge per electron (q):

$$\bar{i} = q\eta\phi_\rho \tag{5.1}$$

where it is assumed that all the photons have sufficient energy to generate free electrons:

$$h\nu > \phi_0 \tag{5.2}$$

where ϕ_0 is the work function of the photocathode. Excess photon energy is converted into kinetic energy of the photoelectron:

$$KE = \tfrac{1}{2}mv^2 = h\nu - \phi_0 \tag{5.3}$$

The photoemissive detector is characterized by high-speed operation (the time between photon absorption and electron collection at the anode can be less than 10^{-10} sec), a large dynamic range, and low noise. The noise is determined primarily by the statistics of photon arrival uncertainty.

Quantum efficiency and spectral response for some common photo-cathodes are shown in Fig. 5-2 and Table 5-1. The semiconductor

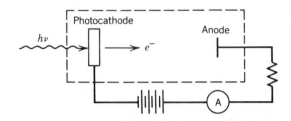

Figure 5-1 Photoemissive detection.

photocathode work functions allow operation at longer wavelengths than is possible with metals. A metal photocathode will usually not respond to photon wavelengths longer than 0.3 μm, and these photocathodes are used as ultraviolet detectors. These ultraviolet detectors are commonly used under vacuum since the atmosphere has very high absorption shortward of 0.28 μm. Fused silica windows for photomultiplier tubes transmit light wavelengths down to about 0.15 μm, while glass cuts off at 0.35 μm. Semiconductor photocathodes, however, respond to wavelengths longer than the wavelength peak due to emission from impurity levels. Therefore an S-20 photocathode, for example, will respond usefully to 0.9 μm even though the peak response is at 0.38 μm (when used with a glass window).

A separate type of semiconducting photocathode that exhibits relatively long-wavelength response at high quantum efficiency is said to have

Table 5-1

Cathode	Composition	λ_p (μm)[a]	η (λ_p)
S-1	AgOCs	0.8	0.004
"S"	Cs_3Sb	0.38	0.16
S-10	BiAgO Cs	0.42	0.068
S-11	Cs_3Sb-O	0.39	0.19
"Super" S-11	CsSb-O	0.41	0.23
S-20	Na_2KSb-Cs	0.38	0.22
Bialkali	K_2CsSb	0.38	0.27
"High Temp" Bialkali	Na_2KSb	0.36	0.21

[a] Assumes that a glass window is used.

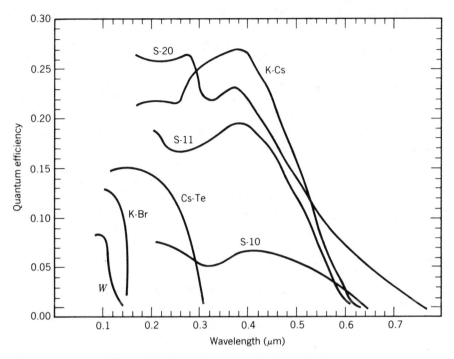

Figure 5-2 Spectral response of photocathodes.

negative electron affinity (NEA). The electron affinity is the energy difference between the vacuum potential and the conduction band:

$$E_a = E_{\text{vac}} - E_c \tag{5.4}$$

Therefore, negative electron affinity implies that any electron in the conduction band will be expelled out of the material (see Fig. 5-3). An NEA photocathode of GaAs-P material can exhibit $\eta > 0.3$ at 0.7 μm.

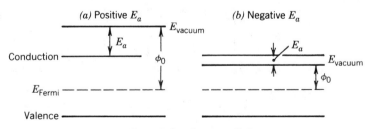

Figure 5-3 Electron affinity.

The current responsivity of a photocathode is

$$\mathcal{R}_i = \frac{\bar{i}}{\phi_e} = \frac{q\eta\phi_p}{\phi_e} = \frac{q\eta[\phi_e/(hc/\lambda)]}{\phi_e} = \frac{q\eta\lambda}{hc} \qquad (5.5)$$

A quantum efficiency of 0.05 at 0.5 μm for an S-10 photocathode would imply a current responsivity of about 0.04 A W^{-1}. The signal current with 100 q sec^{-1} of 0.5 μm wavelength incident on this photocathode will produce a signal current of only $\eta q\phi_p$ or $(0.05)(1.6 \times 10^{-19})(100)$, which is 8×10^{-19} A. The best electrometers can only measure currents that are larger than 10^{-15} A. Therefore, a higher signal current will be required before the detector can be used near its own limits, and not the limits imposed by the measurement electronics. The photomultiplier tube discussed in the next section can produce the electron multiplication (gain) necessary to make photoemissive detectors practical low-light-level detectors.

5-2 PHOTOMULTIPLIER TUBES

A photomultiplier tube (PMT) is illustrated schematically in Fig. 5-4. The PMT consists of a photoemissive detector and a low-noise amplifier contained in the same vacuum jacket. When an incident photon is absorbed by the photocathode to produce a photoelectron, the electron is accelerated toward the first dynode by the voltage drop across R_1. If this potential is 100 V, then the electron will have a kinetic energy of 100 eV

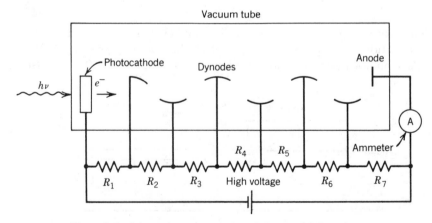

Figure 5-4 Schematic representation of photomultiplier tube.

plus the excess of the photon energy over the work function of the photocathode. This kinetic energy will cause a number of electrons to be emitted from the first dynode. These electrons will each be accelerated toward the second dynode by the voltage drop across R_2. In this way a large number of electrons are collected at the anode for each electron emitted by the photocathode. If three electrons are emitted by each dynode for one incident electron, the gain of the six dynodes is $3^6 = 729$. It is more likely that 11 dynodes would be used in actual practice for a total gain of $3^{11} \simeq 2 \times 10^5$. This gain would be sufficient to raise the photocathode current of 8×10^{-19} A calculated in the example of the previous section to 1.6×10^{-13} A. This signal level is compatible with electronic ammeters.

The gain of a PMT is a function of the applied voltage. If the dynode work function is 2 eV and the incident electron has a kinetic energy of 100 eV, a maximum of 50 secondary electrons could be emitted. The upper limit to secondary electron emission is therefore a linear function of applied voltage. The actual number of secondary electrons that are emitted is usually much less than this upper limit due to penetration of accelerated electrons into the dynode. If this penetration is very deep (more than several atoms), the likelihood of secondary electrons escaping a classical dynode becomes low. An optimum bias voltage can, therefore, be found that trades off electron energy with dynode penetration to achieve maximum gain (see Fig. 5-5). A NEA dynode is not very likely to trap secondary electrons once these electrons are excited to the conduction band, however. Therefore NEA dynodes show nearly linear increase in gain with applied voltage.

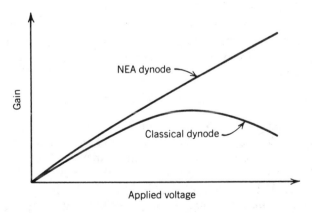

Figure 5-5 Gain of a single dynode as a function of applied voltage.

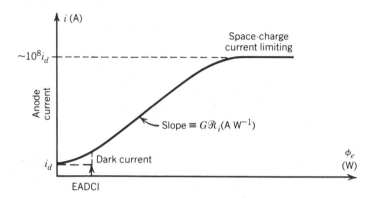

Figure 5-6 PMT linear dynamic range.

This dependence of gain on voltage drop across the interelement resistors of Fig. 5-4 means that the gain is a function of the signal current which flows through these resistors. Gain that is, for most purposes, essentially independent of signal current can be obtained by providing a bias supply current through the resistors which is more than 10 times as high as the largest anode current for which the PMT is rated.

The linear dynamic range of a PMT is typically 10^8 limited by dark current at the low end of the range and space-charge effects at the high end (Fig. 5-6). Space-charge limiting of current occurs when the electron density between the last dynode and the anode becomes so high that the negative space charge due to the electron cloud significantly repels electrons leaving the dynode. The dark current is due to contributions from leakage current (insulator resistance is not quite infinite), regenerative effects (including α, β, and γ particles which cause secondary electron emission), and the thermal (thermionic) emission of electrons. The thermionic emission is often the dominant source of dark current. Richardson's equation yields the magnitude of the photocathode thermionic current:

$$i_T = A_d S T^2 e^{-\phi_0/kT} \tag{5.6}$$

where A_d is the photocathode area, S is a constant ($S = 4\pi m q k^2/h^3$), T is the photocathode temperature, and ϕ_0 is the photocathode work function. The thermionic currents due to the dynodes can also be calculated, but the photocathode usually dominates because this current receives the most electron multiplication before reaching the anode. In any case, the T^2 term in Eq. (5.6) indicates that cooling the PMT will reduce dark current and therefore increase the linear dynamic range at the small-signal end.

Cooling a PMT to about −40°C (233 K) will often reduce the thermionic contribution below the other sources of dark current. A special figure of merit applied only to PMTs (and occasionally to microchannels, which are described in the next section) is the equivalent anode dark current input (EADCI). The EADCI is the value of radiant flux (ϕ_e) incident on the photocathode that will produce an anode current equal to the dark current.

The photomultiplier tube is in effect a vacuum diode and, therefore, shot noise is produced by the variation in the flow of photoelectrons. The shot noise produced at the photocathode is

$$\Delta i_{rms}^K = \sqrt{2q(\bar{i} + i_d)\,\Delta f} \tag{5.7}$$

where i_d is the photocathode dark current and $\bar{i}$ is the cathode photocurrent from Eq. (5.1). The noise equivalent power is usually determined by the dark current:

$$\bar{i}_{NEP} = \Delta i_{rms}^K \tag{5.8}$$

therefore

$$\eta q \phi_p = \eta q \phi_e \frac{\lambda}{hc} = [2qi_d\,\Delta f]^{1/2} \tag{5.9}$$

and we may solve for the NEP:

$$\text{NEP} = \phi_e = \frac{hc}{\eta q \lambda}[2qi_d\Delta f]^{1/2} \tag{5.10}$$

This expression for the NEP assumes that performance is shot noise limited by the dark current.

The total noise at the anode of a PMT includes the effect of gain on the shot noise, and the Johnson noise due the effective resistance of the voltage-dropping resistors:

$$\Delta i_{rms}^A = \sqrt{2qG^2\,(\bar{i} + i_d)\,\Delta f + \frac{4kT\,\Delta f}{R_{eff}}} \tag{5.11}$$

Expression (5.11) indicates that for appropriately high values of the gain (G) and the resistor chain (R_{eff}), a PMT will be shot noise limited. The value of R_{eff} cannot be so high that anode signal current will significantly effect the gain, but the value should be as high as is practical in order to reduce Johnson noise. In practice, it is usually possible to find a PMT that is shot noise limited for the desired application. The gain does not effect the anode signal-to-noise ratio if the PMT is shot noise limited and the gain is constant.

$$\text{SNR}^A = \frac{G\bar{i}}{\Delta i_{rms}^A} = \frac{G\bar{i}}{\sqrt{2qG^2\,(\bar{i}+i_d)\,\Delta f}} = \frac{\bar{i}}{\sqrt{2q(\bar{i}+i_d)\,\Delta f}} \qquad (5.12)$$

Equation (5.12) states that the SNR of a shot-noise-limited PMT is the same at the anode as in the current from the photocathode. The SNR in the current from the photocathode in the absence of dark current is the SNR in the incident photon stream degraded by less than perfectly quantum efficient generation of photelectrons. From Section 1-5, the incident photon SNR is [see Eq. (1.53)]

$$\text{SNR}_{in} = \sqrt{\bar{n}} = \sqrt{\bar{\phi}_p \tau} \qquad (5.13)$$

where $\bar{n}$ is the average number of photons incident in a time interval τ. The degradation due to a quantum efficiency of less than 1 is a Bernoulli process [Eq. (1.50)], and the resulting SNR in the photocurrent is [see Eq. (1.54)]

$$\text{SNR}_{out} = \sqrt{\eta\bar{n}} = \sqrt{\eta}\,\text{SNR}_{in} \qquad (5.14)$$

We conclude this section with a demonstration that signal-dependent noise generated by the dynodes is negligible provided that the first dynode has a large gain (Engstrom, 1980). The first dynode is normally operated with a higher voltage difference with respect to the photocathode than the voltage difference between succeeding dynodes. This typically produces a gain of 5–25 for the first dynode with a gain of about 3 for each following dynode. Starting with an input signal-to-noise ratio of $\sqrt{\bar{n}}$, the SNR after a photocathode of quantum efficiency η is

$$\text{SNR}_{pc} = \frac{\eta\bar{n}}{\sqrt{(\eta\sqrt{\bar{n}})^2 + \bar{n}(1-\eta)\eta}} = \sqrt{\eta\bar{n}} \qquad (5.15)$$

The signal-dependent noise in the number of secondary electrons emitted for each incident electron at the first dynode is σ_1. Then the SNR for a first dynode gain of δ_1 is

$$\text{SNR}_1 = \frac{\eta\bar{n}\,\delta_1}{\sqrt{(\delta_1\sqrt{\eta\bar{n}})^2 + \eta\bar{n}\sigma_1^2}} = \frac{\delta_1\sqrt{\eta\bar{n}}}{\sqrt{\delta_1^2 + \sigma_1^2}} \qquad (5.16)$$

Similarly for the second dynode,

$$\text{SNR}_2 = \frac{\eta\bar{n}\delta_1\delta_2}{\sqrt{\eta\bar{n}(\delta_1^2 + \sigma_1^2)\delta_2^2 + \eta\bar{n}\delta_1\sigma_2^2}} = \frac{\delta_1\delta_2\sqrt{\eta\bar{n}}}{\sqrt{\delta_1^2\delta_2^2 + \sigma_1^2\delta_2^2 + \delta_1\sigma_2^2}} \qquad (5.17)$$

and for k dynodes

$$SNR_k = \frac{\eta\bar{n}\delta_1\delta_2\delta_3\cdots\delta_k}{\delta_1\delta_2\delta_3\cdots\delta_k\sqrt{\eta\bar{n}\left(1+\dfrac{\sigma_1^2}{\delta_1^2}+\dfrac{\sigma_2^2}{\delta_1\delta_2^2}+\dfrac{\sigma_3^2}{\delta_1\delta_2\delta_3^2}+\cdots+\dfrac{\sigma_k^2}{\delta_1\delta_2\delta_3\cdots\delta_k^2}\right)}}$$

$$= \frac{\sqrt{\eta\bar{n}}}{\sqrt{1+\dfrac{1}{\delta_1}\left(\dfrac{\sigma_1^2}{\delta_1}+\dfrac{\sigma_2^2}{\delta_2^2}+\dfrac{\sigma_3^2}{\delta_2\delta_3^2}+\cdots+\dfrac{\sigma_k^2}{\delta_2\delta_3\cdots\delta_k^2}\right)}} \qquad (5.18)$$

We now assume that the secondary electron emission is Poisson distributed. That is, the average number of secondary electrons per incident electron δ_1 is related to the variance in secondary electron emission:

$$\delta_i = \sigma_i^2 \qquad (5.19)$$

Let us also assume that the average dynode gains for $i = 2,3,4\ldots k$ are all equal to δ. The SNR for k dynodes is now

$$SNR_k = \frac{\sqrt{\eta\bar{n}}}{\sqrt{1+\dfrac{1}{\delta_1}\left(1+\dfrac{1}{\delta}+\dfrac{1}{\delta^2}+\dfrac{1}{\delta^3}+\cdots+\dfrac{1}{\delta^{k-1}}\right)}} \qquad (5.20)$$

It is clear that large dynode gains would ensure that there will be a small contribution to noise from the secondary electron emission process. It will be more clear that the first dynode dominates the noise contribution of the amplification process by rewriting expression (5.20) making use of the identity

$$\frac{1}{1-x} = \sum_{i=0}^{\infty} x^i \qquad (5.21)$$

Using this fact in Eq. (5.20),

$$SNR_k \approx \frac{\sqrt{\eta\bar{n}}}{\sqrt{1+\dfrac{1}{\delta_1}\left(\dfrac{1}{1-1/\delta}\right)}} = \frac{\sqrt{\eta\bar{n}}}{\sqrt{1+\dfrac{1}{\delta_1}\left(\dfrac{\delta}{\delta-1}\right)}} \qquad (5.22)$$

For large δ_1, the dynode chain amplification is noiseless, that is,

$$SNR_k = \sqrt{\eta\bar{n}} = SNR_{photocathode} \qquad (5.23)$$

The first dynode gain should be as large as is practical because this gain dominates the noise characteristics of the entire photomultiplier tube.

5-3   MICROCHANNELS

Microchannels are small tubes of glass that operate as electron multipliers (Fig. 5-7). The advantages of small size and relative ease of manufacture make microchannels adaptable to imaging applications and, for applications that require immunity from large magnetic fields, they are alternatives to photomultiplier tubes. Glass tubes made of a combination of oxides of silicon, lead, and alkali compounds in mixture have been used to obtain the desired resistivity in the approximate range of 10^{10}–$10^{15} \, \Omega \, \mathrm{cm}^{-2}$ while providing a gain of two electrons ejected per single wall collision.

Single microchannels have been constructed with lengths on the order of 5 cm. These long channels yield electron gains as high as 10^8 with an applied potential approaching 3000 V. The channel aperture can be increased to exceed 1 cm in diameter by forming a glass cone at the end of

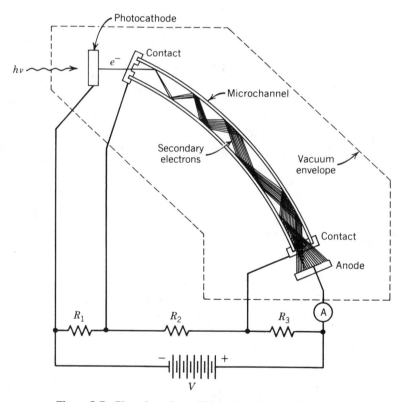

Figure 5-7   Photodetection utilizing microchannel electron gain.

the channel nearest the photocathode. As an optical radiation detector, this device is limited to very low light levels before the small-diameter channel reaches current saturation in the neighborhood of 1 μA for large 1-mm-diameter channels. The dark current for such a channel is typically 10 pA, yielding a dynamic range of only 10^5. The photomultiplier tube therefore offers superior performance in most single-detector applications. The principal advantage of microchannels is that they can be constructed in arrays for imaging applications.

5-4 MICROCHANNEL PLATES

A microchannel plate is an array of microchannels formed in a single plate of glass. Typical physical dimensions for such an array include an overall diameter of 18–75 mm, a thickness (channel length) of about 1 mm, and channel diameters in the range 8–20 μm. A total of 10^4–10^7 microchannels may be present in a single plate. Each microchannel can be used to produce one intensified picture element ("pixel") in an imaging system.

Microchannel plate gain is much less than that of many photomultiplier tubes or long single channels, because the channels are relatively short. Gain uniformity from channel to channel across the plate is a function of channel-diameter uniformity. Channel-diameter uniformity of about 2% and hence 4 or 5% gain uniformity is commonly achieved. The electron gain of a single microchannel plate is usually limited to about 10^4, although two or more plates can be cascaded to increase the gain. When two plates are used in series (Fig. 5-8), the gain may be increased to a maximum of about 10^7. The spatial resolution is usually reduced by a factor of 2 because of the spread of electrons leaving the first plate. This results in a finite probability of exciting channels adjacent to the intended channel in the second plate. The channels are usually inclined by 5° in a microchannel plate (exaggerated in Fig. 5-8), and the inclines are opposed in a chevron pattern when two plates are cascaded. The purpose of this is to reduce the ion feedback that may occur when residual gas molecules within the vacuum envelope are ionized by the accelerated electrons. These positive ions are attracted backward up the channel to the negative potential end where they may initiate new electron multiplication chains. Ion feedback, at best, creates false signal levels. At worst, physical damage to the microchannel plates or the anode may occur due to high current densities. A current density of several microamperes per square centimeter of plate will probably cause permanent damage. Space charge may limit current density before damage can occur. A linear space-charge density of 10^6–10^8 electrons per millimeter will balance the 1–3 kV bias

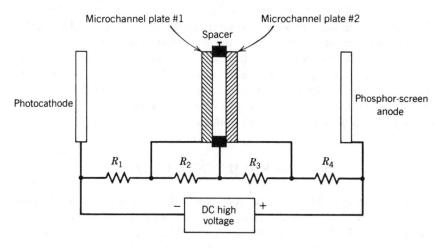

Figure 5-8   Image intensifier using two microchannel plates.

applied across the microchannel plates and overall electron gain is therefore limited to about 10^8 for a single device no matter how many plates are cascaded. Figure 5-9 illustrates how space-charge effects can provide an "automatic-gain-control" action.

The microchannel plate offers the possibility of imaging with a signal-to-noise ratio that approaches the quantum noise limit imposed by photon statistics and the quantum efficiency of the particular photocathode used:

$$\text{SNR} \le \sqrt{\eta \bar{N}} \qquad (5.24)$$

where η is the photocathode quantum efficiency and $\bar{N}$ is the average number of incident photons in one integration time. The equality of Eq. (5.15) holds for a pixel the size of one channel, if we ignore imperfect electron beam collimation. The signal-to-noise ratio at the output of a device of the type illustrated in Fig. 5-8 will be less than $\sqrt{\eta \bar{N}}$, however, when integrated over the entire active area of the device. This is because there must be solid glass between the channels where electron collection and multiplication (gain) does not occur. If the device of Fig. 5-8 is uniformly irradiated over its photocathode with $\bar{N}$ photons total per integration time, the output signal-to-noise ratio is limited to

$$\text{SNR} = \sqrt{\eta F_f \bar{N}} = \frac{\sqrt{\eta \bar{N}}}{F_n} \qquad (5.25)$$

where F_f is the filling factor or fraction of microchannel plate area that

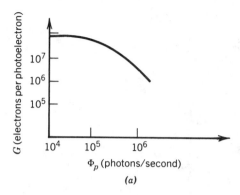

(a)

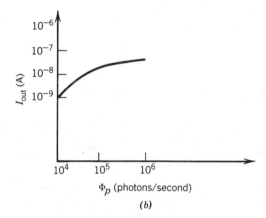

(b)

Figure 5-9 Gain limitation at high input photon flux: (a) Gain decline due to space charge; (b) output saturation.

collects photoelectrons and F_n is the noise factor or the factor by which the noise appears to increase as a result of the filling factor being less than 1. The noise figure (in decibels) is related to the noise factor (NF):

$$\text{NF} = 20 \log_{10}(F_n) = 20 \log_{10} \frac{1}{\sqrt{F_f}} = 10 \log_{10} \frac{1}{F_f} \qquad (5.26)$$

A typical microchannel plate may have $f_f = 0.60$, that is 60% of the plate area collects photoelectrons. The corresponding noise factor f_n is about 1.3. The SNR is therefore less than 80% of what it would be if the fill factor was 1.0. The noise figure is about 2 dB, that is, the SNR is about

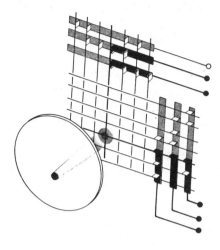

2 dB lower than it would be for a "perfect" microchannel plate. Some advanced microchannel plates have a fill factor of 0.8, so that the SNR is 90% of ideal, or down about 1.4 dB from ideal. This high fill factor is achieved by beveling the channel entrance to increase the collecting area.

Electronic readout of the intensified electron image generated by the microchannel plate is often desired. It is possible, in principle, to provide a separate metal anode for each channel, but this method is not practical for the large number of channels available in some plates (10^7). If fast electronics are used, the very fast response time of the microchannel plate (10^{-9} sec) allows pulse counting up to relatively high incident photon fluxes (perhaps 10^{10} photons sec^{-1} depending on the quantum efficiency of the photocathode). With only one pulse at a time to locate spatially, efficient methods of encoding the pulse position are possible. In Fig. 5-10, only six wires are used to encode 49 positions. this method may be extended so that a given number of connections, N_c, can encode the position of a larger number of positions N_p:

$$N_p = (2^{N_c/2} - 1)^2 \tag{5.27}$$

If the speed of the electronics is not sufficient to separate pulses, a method that stores the image for later readout must be used. A charge transfer device (see Chapter 9) may be used, or photographic film may be used for later development and digitization in non-real-time applications.

5-5   INTRODUCTION TO IMAGING SYSTEM CONSIDERATIONS

This section will compare an imaging system using a photomultiplier tube with an imaging system using a microchannel plate (Fig. 5-11). The advantage of using a detector array for imaging should then be clear. The comparison between discrete detectors and array detectors for imaging sets the scope of this section.

The device in Fig. 5-11a forms an image by scanning a field of view with vibrating and rotating mirrors. Only a single resolution element is present on the PMT detector at any one time. If the full image frame is divided into N picture elements (pixels), and the time required to scan a full frame is T_{frame}, then the dwell time on a single pixel is

$$\tau_d = \frac{T_{\text{frame}}}{N} \tag{5.28}$$

The electrical bandwidth required at the output of the PMT is then

$$\Delta f = \frac{1}{2\tau_d} \tag{5.29}$$

The array imaging system of Fig. 5-11b has the advantage that each pixel is being detected for a full frame time:

$$\tau_d = T_{\text{frame}} \tag{5.30}$$

The ratio of the bandwidth required for the array (Δf_a) to the bandwidth

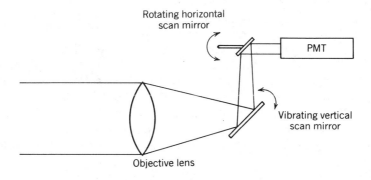

Figure 5-11a   Scanning-photomultiplier-tube imager.

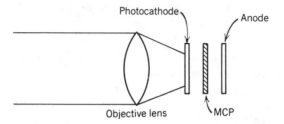

Figure 5-11b   Microchannel-plate-array imager.

required for the scanner (Δf_s) is

$$\frac{\Delta f_a}{\Delta f_s} = \frac{1/2\,T_{\text{frame}}}{N/2\,T_{\text{frame}}} = \frac{1}{N} \qquad (5.31)$$

It is now possible to compare the performance of the two imaging systems. If the detectivity per pixel and pixel area are the same for both devices, we can compare the signal-to-noise ratios produced when both devices image the same scene brightness:

$$\frac{\text{SNR}_a}{\text{SNR}_s} = \frac{D^*\Phi/\sqrt{A_d \Delta f_a}}{D^*\Phi/\sqrt{A_d \Delta f_s}} = \sqrt{\frac{\Delta f_s}{\Delta f_a}} = \sqrt{N} \qquad (5.32)$$

It can therefore be seen that an SNR advantage of $\sqrt{N}$ accrues to an array imager compared to a scanning imager using a single-detector element. In a system of practical interest, the number of pixels can reach well above 500×500. The SNR advantage of $\sqrt{500^2} = 500$ is very large. This motivates the development of array imaging systems based on microchannel plates (Section 5-4) and charge transfer devices (Chapter 9).

The imaging system of Fig. 5-12 will be briefly analyzed to illustrate the performance achievable with modern components. A photocathode produces photoelectrons when illuminated. The photoelectron flux is amplified preserving spatial (image) information using two microchannel plates in cascade. The photoelectrons that exit the second microchannel plate are accelerated toward the phosphor where photons are produced by the photoelectron kinetic energy. The light is guided into a CCD detector array using a fiber-optic plate. We ignore here noise sources of the CCD array, which are discussed in Chapter 9. The signal-to-noise ratio for this system is then

$$\text{SNR} = \frac{\bar{n}\eta_{\text{pc}} F_{M1} G_{M1} \eta_{M1} F_{M2} G_{M2}\ \eta_{M2} G_{\text{phos}} \eta_{\text{phos}} F_{\text{fo}} F_{\text{CCD}} \eta_{\text{CCD}}}{\sqrt{\sigma_{\text{pc}}^2 + \sigma_{M1}^2 + \sigma_{M2}^2 + \sigma_{\text{phos}}^2 + \sigma_{\text{fo}}^2 + \sigma_{\text{CCD}}^2}} \qquad (5.33)$$

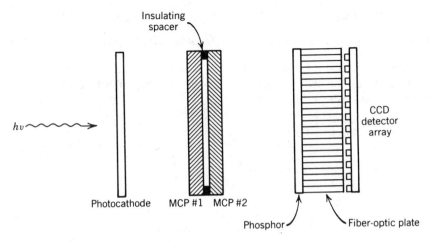

Figure 5-12   Array-imaging-system components.

where $\bar{n}$ = average number of incident photons,

η_{pc} = photocathode quantum efficiency,

F_{M1} = fill factor of the first multichannel plate MCP,

G_{M1} = gain of the first MCP,

η_{M1} = quantum efficiency of the first MCP,

F_{M2} = fill factor of the second MCP,

G_{M2} = gain of the second MCP,

η_{M2} = quantum efficiency of the second MCP,

G_{phos} = phosphor gain due to the photoelectron kinetic energy being large enough to produce multiple photons per incident photo-electron,

η_{phos} = phosphor quantum efficiency,

F_{fo} = fibre-optic plate fill factor,

F_{CCD} = CCD fill factor,

η_{CCD} = CCD quantum efficiency,

The noise variances are all expressed at the appropriate output levels and assumed to be independent Poisson processes:

$$\sigma_{pc}^2 = \bar{n}\eta_{pc}F_{M1}[\,G_{M1}\,\eta_{M1}F_{M2}\,G_{M2}\,\eta_{M2}\,G_{phos}\,\eta_{phos}F_{fo}F_{CCD}\,\eta_{CCD}]^2 \qquad (5.34)$$

$$\sigma_{M1}^2 = \bar{n}\eta_{pc}F_{M1}\,G_{M1}\,\eta_{M1}[\,F_{M2}\,G_{M2}\,\eta_{M2}\,G_{phos}\,\eta_{phos}F_{fo}F_{CCD}\,\eta_{CCD}]^2 \qquad (5.35)$$

$$\sigma_{M2}^2 = \bar{n}\eta_{pc}F_{M1}\,G_{M1}\,\eta_{M1}F_{M2}\,G_{M2}\,\eta_{M2}[\,G_{phos}\,\eta_{phos}F_{fo}F_{CCD}\,\eta_{CCD}]^2 \qquad (5.36)$$

$$\sigma_{phos}^2 = \bar{n}\eta_{pc}F_{M1}\,G_{M1}\,\eta_{M1}F_{M2}\,G_{M2}\,\eta_{M2}\,G_{phos}\,\eta_{phos}[\,F_{fo}F_{CCD}\,\eta_{CCD}]^2 \qquad (5.37)$$

$$\sigma_{fo}^2 = \bar{n}\eta_{pc}F_{M1}G_{M1}\eta_{M1}F_{M2}G_{M2}\eta_{M2}G_{phos}\eta_{phos}F_{fo}[F_{CCD}\eta_{CCD}]^2 \qquad (5.38)$$

$$\sigma_{CCD}^2 = \bar{n}\eta_{pc}F_{M1}G_{M1}\eta_{M1}F_{M2}G_{M2}\eta_{M2}G_{phos}\eta_{phos}F_{fo}F_{CCD}\eta_{CCD} \qquad (5.39)$$

Substituting Eq. (5.34) through (5.39) into Eq. (5.33) and setting the parameters of the two microchannel plates equal (subscript $M1$ =subscript $M2$ = subscript M):

$$SNR = \sqrt{\bar{n}\eta_{pc}F_M}\left[1 + \frac{1}{G_M\eta_M} + \frac{1}{F_M(G_M\eta_M)^2} + \frac{1}{F_M(G_M\eta_M)^2 G_{phos}\eta_{phos}} \right.$$

$$+ \frac{1}{F_M(G_M\eta_M)^2 G_{phos}\eta_{phos}F_{fo}}$$

$$\left. + \frac{1}{F_M(G_M\eta_M)^2 G_{phos}\eta_{phos}F_{fo}F_{CCD}\eta_{CCD}} \right]^{-1/2} \qquad (5.40)$$

For some typical parameter values $F_M = 0.8$, $G_M = 10^4$, $\eta_M \approx 1$, $G_{phos} = 100$, $\eta_{phos} = 0.5$, $F_{fo} = 0.5$, $F_{CCD} = 0.25$, and $\eta_{CCD} = 0.5$, it is clear that Eq. (5.40) is essentially

$$SNR = \sqrt{\bar{n}\eta_{pc}F_M} = \sqrt{\eta_{pc}F_M}\,(SNR)_{photon\ limit} \qquad (5.41)$$

BIBLIOGRAPHY

Engstrom, R. W., *Photomultiplier Handbook*, RCA, 1980.

Hennes, J. and L. Dunkelman, "Ultraviolet Technology," in *The Middle Ultraviolet: Its Science and Technology*, A. E. S. Green, Ed., Wiley, New York, 1966, Chap. 15.

Jacobs, S. F., "Nonimaging Detectors," in *Handbook of Optics*, W. G. Driscoll and W. Vaughan, Eds., McGraw-Hill, New York, 1978, Section 4.

Lampton, M., "The Microchannel Image Intensifier," *Scientific American*, Nov. 1981.

Martinelli, R. V. and D. G. Fisher, "The Appplication of Semiconductors with Negative Electron Affinity Surfaces to Electron Emission Devices," *Proc. IEEE* **62** (10), October 1974.

Wiza, J. L., "The Microchannel Plate," *Optical Spectra*, April, 1981.

PROBLEMS

5-1.   Discuss how the signal-to noise ratio of a photomultiplier tube (PMT) varies with:

    a. Overall gain.

    b. Applied voltage.

5-2. Discuss the factors that limit the ultimate speed of response of a PMT.

5-3. Discuss various dynode geometrical configurations of a PMT.

5-4. Compare a transmission-mode and a reflection-mode photocathode.

5-5. Describe the dependence of dark current on temperature for a PMT (especially at temperatures below $-50°C$).

5-6. What factors limit the ultimate current of a PMT?

5-7. Discuss the negative-electron-affinity photocathode.

5-8. What factors limit the dynamic range of a PMT?

5-9. You are given a 1 cm^2 detector and told it has an S-20 response and a sensitivity of 1 A/lumen from a 2870 K blackbody source. What is the output of the detector when exposed to:

 a. Sunlight (100 W m^{-2}, 6000 K).

 b. HeNe laser (1 mW, 632.8 nm).

 c. GaAs LED (10 mW, 930 nm).

5-10. What is limiting noise of PMT? How would you minimize it?

5-11. Calculate the signal shot noise from the arrival rate of $N = 10^5$ photons /sec^{-1} at the input of the PMT and the quantum efficiency of 20% ($\eta = 0.2$).

5-12. The electron multiplication occurring from dynode to dynode in the multiplier section of a PMT can be considered as a series of events in cascade. The PMT has a total of n dynodes, each one having a mean secondary electron emission coefficient ("gain") δ_i:

 a. Calculate the mean number of electrons N_e ("overall gain") produced at the output for a single electron at the input.

 b. Calculate the variance for N_e as a function of the variances of the individual dynodes assuming that they are statistically independent. In order to obtain a simple expression assume that all gains except that for the first stage are the same and these variances including that for the first stage follow Poisson statistics.

 c. Assuming you have during a certain observation time at the input to the PMT multiplier section n_e photoelectrons with a standard deviation $\delta = \sqrt{n_e}$ such that your signal-to-noise ratio is $SNR_{in} = \sqrt{n_e}$. What is the signal-to-noise ratio SNR_{out} at the output of the PMT (gain of the first stage is δ_1, the gain of each of the following stages is δ). How do you have to choose the gain of the first stage in order to achieve noiseless amplification by the PMT?

5-13. Derive the expressions for the mean and variance of the binomial distribution. The binomial probability distribution function is

$$P(r) = \binom{n}{r} p^r q^{n-r} \qquad \binom{n}{r} = \frac{n!}{r!(n-r)!}$$

where $n \equiv$ number of trails,

$r \equiv$ number of events in n trials,

$p \equiv$ probability of an event occurring,

$q = 1 - p \equiv$ probability an event does not occur.

5-14. You are given a 1 cm² detector and told it has an S-11 response and a sensitivity of 25 μA/lumen from a 2856 K blackbody source. What is the output of the detector when exposed to (use Fig. 5.2):

a.   HeNe laser (1 mW, 632.8 nm).

b.   GaAs LED (10 mW, 930 nm).

5-15. The electron multiplication occurring from dynode to dynode in the multiplier section of the PMT can be considered to a first approximation as a series of events in cascade. The PMT has a total of n dynodes each one having a mean secondary electron emission coefficient of $\bar{\delta_i}$ and variance of σ_i^2; and a mean collection efficiency of η_c.

a.   Calculate the mean number of electrons $\bar{N}$ produced at the output for a single electron at the input.

b.   Develop a formula for the variance of the gain of the multiplier as a function of the relative variances of the individual dynodes and collection efficiency.

What are the conditions for which only the variance of the first dynode is of importance?

CHAPTER 6

THERMAL DETECTORS AND THERMOPILES

6-1 INTRODUCTION

Thermal detectors are the oldest man-made detectors for sensing radiation. The first thermal detector can probably be credited to Sir William Herschel when he used a thermometer past the red end of the spectrum produced by a prism to verify the presence of infrared radiation in 1800. However, in about the same time frame, Seebeck in 1825 discovered the thermocouple. Nobili used the thermocouple to make a thermopile in 1829, and, in 1880, Langley developed the bolometer as a radiation detector.

Most thermal detectors operate at room temperature and have a large region of spectral response. Thermal detectors absorb radiation, which produces a temperature change that in turn changes a physical or electrical property of the detector. Since a change in temperature takes place, the thermal detectors are inherently slow responding and have relatively low sensitivity compared to other detectors.

Some comparisons can be made between thermal detectors and photodetectors, as shown in Table 6-1.

In previous chapters, we have discussed photodetectors in which radiation of sufficiently short wavelength interacts directly with the lattice sites giving rise to excess current carriers. This produces a current, voltage, or resistance change. The spectral response depends on the energy gap of the semiconductor used as the detector.

For thermal detectors, a temperature change is produced in the

Table 6-1

General Properties of Thermal and Photodetectors

Parameter	Thermal	Photodetector
Frequency response	Low	High
Spectral responsivity	Wide—constant	Limited—λ dependent
Sensitivity	Low	High
Operating temperature	Room	Cryogenic
Cost	Economical	Relatively expensive

detector by the amount of power that is absorbed by the detector. Therefore, the spectral absorptance of the detector determines the spectral response. The absorptance is often a function of wavelength. In addition, since the heat capacity or thermal mass influences the amount of temperature change, the detectors are characteristically small in volume in order to have small heat capacity and a faster time response.

Thermal detectors are commonly used where spectrally flat responsivity is important, for example, in calibration standards.

Some types of thermal detectors are:

1. *Mechanical Displacement*
   Liquid in a glass thermometer
   Bimetal strip
   Crooke's radiometer
   Golay Cell
2. *Electrical*
   Bolometer
   Thermocouple (thermopile)
   Pyroelectric
3. *Other*
   Evapograph

All thermal detectors can be thought of as having two essential parts that make up the device: the absorber and the temperature transducer (temperature sensor). Sometimes one material acts as both; however, more commonly, a blackening substance is placed over the temperature sensor.

This chapter will review the heat-balance equation for a thermal detector and the background-limited performance of such a device, and discuss the thermocouple and thermopile. The bolometer and pyroelectric detector will be discussed in subsequent chapters.

6-2 HEAT BALANCE EQUATION

The thermal detector works via the heating of the detector chip; therefore, a heat equation needs to be examined to model mathematically the thermal detector.

Consider a detector element suspended on two wires, which are connected to the heat sink as shown in Fig. 6-1. The heat sink is at some ambient temperature T_a.

One can write an equation for the rate of heat exchange between the detector and the heat sink. The heat input to the detector via radiant energy (ϕ_e) is used to warm up the detector chip. The loss mechanisms are due to conduction down the leads and/or radiant emission of the detector itself. The expression for warming of the detector chip is

$$H \frac{dT}{dt} = C_p m \frac{dT}{dt} \tag{6.1}$$

where H = heat capacity (J K^{-1}),
$\quad C_p$ = specific heat,
$\quad m$ = mass density of the detector.

The heat capacity of the detector is

$$H = C_p m \tag{6.2}$$

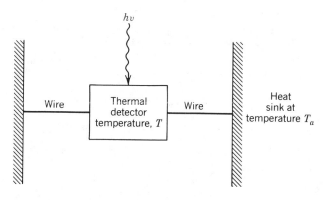

Figure 6-1 Thermal detector mounted via wires to heat sink.

The radiative heat loss is

$$A_d \sigma [T^4 - T_a^4] = \text{heat loss due to radiation} \tag{6.3}$$

where A_d = detector area,
σ = Stefan–Boltzmann constant,
T = detector temperature.

The detector temperature, however, is very near the ambient temperature, so the radiation heat loss is negligible for most situations.

The heat loss due to thermal conduction down the wires is

$$K \, dT = \text{heat loss due to conduction} \tag{6.4}$$

where K = thermal conductance (W deg^{-1}).

If we set up the heat-balance equation, we obtain

$$H \frac{d\Delta T}{dt} + K \Delta T = \phi_e \tag{6.5}$$

Assuming the radiant power to be a periodic function,

$$\phi_e = \phi_{e0} e^{i\omega t} \tag{6.6}$$

where ϕ_{e0} = amplitude of sinusoidal radiation,
$\omega = 2\pi f$.

The solution to the differential heat equation is

$$\Delta T = \Delta T_0 e^{-K/H\omega t} + \frac{\phi_{e0} e^{j\omega t}}{K + j\omega H} \tag{6.7}$$

The first term is the transient part, and, as time increases, this term exponentially decreases to zero, so it can be dropped with no loss of generality for the change in temperature. Therefore, the change in temperature of any thermal detector due to incident radiative flux is (assuming a detector emissivity ε)

$$\Delta T = \frac{2\phi_{e0} \varepsilon}{K[1 + \omega^2 H^2/K^2]^{1/2}} \tag{6.8}$$

From the form of this transfer equation, a thermal time constant (τ_T) can be defined

$$\tau_T = \frac{H}{K} \tag{6.9}$$

This time constant determines the speed of response of thermal detectors.

If the wires have high thermal conductivity (large cross section, good thermally conducting material), the detector will respond very quickly. If, however, the detector has a large heat capacity, its temporal response will be slow.

Typical values of thermal time constants are in the millisecond range. There is a trade-off between sensitivity, ΔT, and frequency response. If one wants a high sensitivity, then a low frequency response is forced upon the detector. Therefore, one cannot have both a high sensitivity and a fast responding detector.

The frequency response of this temperature change is shown in Fig. 6-2. The 3-dB break point is determined by the thermal time constant:

$$f_c = \frac{1}{2\pi\tau_T} \tag{6.10}$$

The change in detector temperature ΔT is produced by incident radiation. If no radiation is incident, the detector temperature should be equal to the heat-sink temperature. However, it will have some fluctuation around this average value and is called temperature noise. This temperature noise sets the limit to sensitivity and determines the minimum radiant power that can be detected. The incident radiation must produce a temperature change comparable to the temperature noise in order to be detected.

The variance in temperature (temperature noise) can be shown to be (Kruse et al., 1963)

$$\overline{\Delta T^2} = \frac{4kKT^2\Delta f}{K^2 + \omega^2 H^2} \tag{6.11}$$

where k is the Boltzmann constant. Obviously, the larger the temperature change ΔT produced from signal radiation [Eq. (6.8)], the larger the signal-to-noise ratio, therefore, in a thermal detector the heat capacity (H) and thermal conductance (K) to the heat sink should be as small as

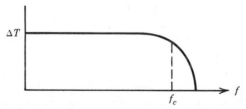

Figure 6-2 Frequency response of the temperature change for a thermal detector.

possible. This means that a small detector mass and fine connecting wires to the heat sink are desirable.

6-3 PHOTON-NOISE-LIMITED PERFORMANCE OF THERMAL DETECTORS

The title of this section is something of a misnomer. The section could be called Background-Radiation-Noise-Limited Performance since thermal detectors respond to radiant power instead of photons. Common usage of the term photon-noise-limited performance throughout the literature compels its use here.

As stated in the previous section, the thermal conductance from the detector to the outside world should be small. The smallest possible thermal conductance would occur when the detector is completely isolated from the environment under vacuum with only radiative heat exchange between it and its heat-sink enclosure. Such an ideal model can give us the ultimate performance limit of a thermal detector. Figure 6-3 shows such a configuration in which the thermal detector is at temperature T_1, the environment is at T_2, and the following assumptions are made:

1. A perfect vacuum surrounds the detector.
2. There are no wire connections to the external environment.
3. The detector temperature is T_1.
4. Environment (background) at temperature T_2 surrounds the detector.
5. Emissivity = absorptivity = ε is constant, independent of the wavelength of radiation or temperature changes.

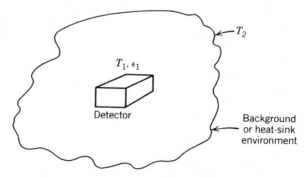

Figure 6-3 Ideal thermal detector completely isolated from heat sink except for radiation exchange.

Consider the radiative exchange between the detector and background to determine the fluctuation in temperature due to the fluctuations in the radiation absorbed. This fluctuation in temperature can be considered to be the radiation-noise-limited case and is related to the fluctuations in the instantaneous number of photons emitted by a radiating source.

Consider the variance in emitted power (σ_ϕ^2) in terms of noise spectral density as (following Kruse et al., 1963)

$$\sigma_\phi^2 = \langle [\phi_e(t) - \langle \phi_e \rangle]^2 \rangle = \int_0^\infty P(f)\,df \tag{6.12}$$

where $P(f) =$ mean square noise power per unit bandwidth

which can be expressed as

$$P(f) = 2(h\nu)^2 \bar{N} \tag{6.13}$$

where $\bar{N}$ is the average emission rate at an energy level, $h\nu$. This is Carson's theorem, and it has been assumed that the spectrum of the noise is white (uniform) (Yariv, 1976).

The average rate of power radiated divided by the energy associated with each emission event is $\bar{N}$, or

$$\bar{N} = \frac{M_e(\nu, t)}{h\nu} \tag{6.14}$$

Substituting Planck's radiation law (Chapter 1)

$$\bar{N} = \frac{2\pi\nu^2/c^2}{e^{h\nu/kT} - 1} \tag{6.15}$$

Then the mean square noise power density per spectral interval is

$$P(f) = \frac{2(h\nu)^2 \cdot 2\pi\nu^2/c^2}{(e^{h\nu/kT} - 1)} \tag{6.16}$$

We can now find the variance in emitted power (σ_ϕ^2) by integration, but, since thermal detectors respond to all wavelengths, we must also include the Bose–Einstein factor in Eq. (6.16):

$$F_{\text{B-E}} = \frac{e^{h\nu/kT}}{e^{h\nu/kT} - 1} \tag{6.17}$$

Therefore

$$P(f) = \int_0^\infty 2(h\nu)^2 \frac{2\pi\nu^2/c^2}{(e^{h\nu/kT} - 1)} \frac{e^{h\nu/kT}}{e^{h\nu/kT} - 1}\,d\nu \tag{6.18}$$

Integrating,

$$P(f) = 8k\sigma T_2^5 \, \text{W m}^{-2} \, \text{Hz}^{-1} \tag{6.19}$$

The mean square noise power on a detector of area A_d and emissivity (absorptivity) of ε, with a bandwidth Δf for white noise is

$$\sigma_{\phi_e}^2 = 8k\sigma T_2^5 A_d \varepsilon \, \Delta f \tag{6.20}$$

This is the noise associated with a blackbody at temperature (T_2).

Since there is also a contribution to the noise due to random power radiated by a detector at temperature T_1, the total noise power is the sum of the variances of absorbed (received) power plus self-emitted radiated power. Assuming the emissivity is the same,

$$\overline{\Delta \phi_e^2} = 8 A_d \varepsilon k\sigma \, \Delta f (T_2^5 + T_1^5) \tag{6.21}$$

the total mean square radiation noise.

To calculate the NEP, the required condition is to set incident power equal to noise power,

$$\varepsilon \phi_e = \left[\overline{\Delta \phi_e^2} \right]^{1/2} \tag{6.22}$$

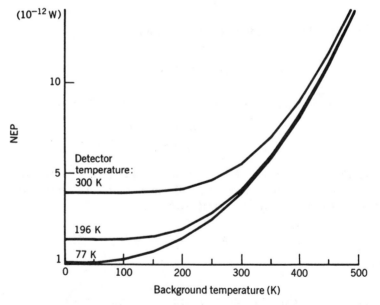

Figure 6-4   NEP versus thermal-detector temperature for radiation-noise-limited operation.

and therefore:

$$\text{NEP} = [8A_d k\sigma \,\Delta f(T_2^5 + T_1^5)]^{1/2}/\sqrt{\varepsilon} \qquad (6.23)$$

This is analogous to NEP_{BLIP} for a photodetector.

As an example, for $A_d = 1 \text{ mm}^2$, $\varepsilon = 1$, $T_1 = T_2 = 300 \text{ K}$, and $\Delta f = 1 \text{ Hz}$, we obtain $\text{NEP} = 5.6 \times 10^{-12} \text{ W Hz}^{-1/2}$. It is interesting to note that if the detector is cooled to liquid-helium temperature or even to absolute zero, the NEP would only improve by a factor of $\sqrt{2}$. Figure 6-4 shows a plot of NEP versus background temperature T_2 for various detector temperatures.

6-4   THERMOCOUPLES

The thermocouple consists of two dissimilar metals connected in series. As the temperature of this junction varies, the electromotive force developed at the output terminals varies. This device is more correctly called the thermoelectric couple, because if two dissimilar metal wires are connected at both ends, and one end is heated, current will flow in the electrical loop.

Figure 6-5 shows two dissimilar metal wires connected in series. The amount of voltage needed to stop current flow at junction 2 is the emf developed by the thermocouple.

In most cases the thermocouple is tied in series to a twin so that a

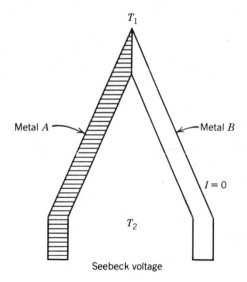

Metal A

Metal B

T_1

T_2

$I = 0$

Seebeck voltage

Figure 6-5 Thermoelectric effect with two dissimilar metals.

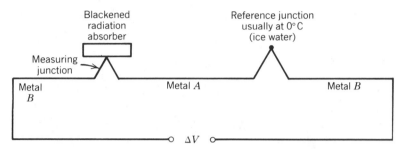

Figure 6-6 Two dissimilar metals connected in series.

reference temperature junction can be established. In the configuration shown in Fig. 6-5 the voltage reading would be referenced to ambient temperature, which is not well defined. A more-common approach is shown in Fig. 6-6 (Travers, 1982). The voltage produced is proportional to the temperature difference

$$\Delta V = \alpha \Delta T \qquad (6.24)$$

where α is the Seebeck coefficient.

The operation of a thermocouple can be partially explained by discussing and quantifying the solid-state physics of a metal. Recall from Chapter 2 that the Fermi energy distribution predicts the probability that an energy level will be occupied by an electron at a given temperature. The higher the energy of an electron, the higher the probability that it will enter the conduction band. If one examines the electron occupancy around the Fermi level for an energy distance of kT on both sides of a junction, the Fermi function is near either zero or one. Therefore, only electrons in the narrow energy band ($E_F \pm kT$) can be considered to be available. Some typical values of Seebeck coefficients are shown in Table 6-2. Table 6-3

Table 6-2

Some Commonly Used Thermocouples

Material	$\alpha \ (\mu V/°C)$
Bi–Sb	100
Fe–constantan	54
Co–constantan	39

shows the thermoelectric voltage produced for copper–constantan versus temperature.

In terms of fractional energy, the electrons that are available are $2kT/E_F$ of the entire population. The total energy of the system is then the number of valence electrons (n) times Avagadro's number (N) times the energy per electron (kT):

$$E_{tot} = nNkT \tag{6.25}$$

But only a fraction of these electrons are activated thermally, so the internal energy produced in a solid due to electrons is

$$U = \frac{2kT}{E_F} nNkT$$

$$= \frac{2nNk^2 T^2}{E_F} \tag{6.26}$$

The electron contribution to specific heat is

$$\frac{\partial U}{\partial T} = \frac{4nNk^2 T}{E_F} \tag{6.27}$$

The voltage produced is therefore approximately

$$\alpha \approx \frac{2k}{eE_F} \tag{6.28}$$

which indicates that the Seebeck coefficient is a function of temperature and Fermi level. The value of about $100~\mu V/K$ at 300 K operation from Eq. (6.28) is in reasonable agreement with Table 6-2.

The heat-balance equations of Section 6-2 showed that the change in detector temperature due to radiation on a thermal detector is

$$\Delta T = \frac{2\phi_e \varepsilon}{[K^2 + \omega^2 H^2]^{1/2}} \tag{6.29}$$

where ϕ_e = amplitude of sinusoidal flux,
ε = detector absorber emissivity,
K = thermal conductance (W deg^{-1}),
H = heat capacity (J deg^{-1}).

and the thermal time constant (τ_T) is

$$\tau_T = \frac{H}{K} \tag{6.30}$$

Table 6-3

Thermoelectric Voltage Produced by Copper–Constantan Junction with a Reference Junction at 0°C (Courtesy of OMEGA Engineering, Inc., an OMEGA Group Company.)

°C	0	1	2	3	4	5	6	7	8	9	10	°C
					Thermoelectric Voltage in Absolute Mill Volts							
-270	-6.258											-270
-260	-6.232	-6.236	-6.239	-6.242	-6.245	-6.248	-6.251	-6.253	-6.255	-6.256	-6.258	-260
-250	-6.181	-6.187	-6.193	-6.198	-6.204	-6.209	-6.214	-6.219	-6.224	-6.228	-6.232	-250
-240	-6.105	-6.114	-6.122	-6.130	-6.138	-6.146	-6.153	-6.160	-6.167	-6.174	-6.181	-240
-230	-6.007	-6.018	-6.028	-6.039	-6.049	-6.059	-6.068	-6.078	-6.087	-6.096	-6.105	-230
-220	-5.889	-5.901	-5.914	-5.926	-5.938	-5.950	-5.962	-5.973	-5.985	-5.996	-6.007	-220
-210	-5.573	-5.767	-5.782	-5.795	-5.809	-5.823	-5.836	-5.850	-5.863	-5.876	-5.889	-210
-200	-5.603	-5.619	-5.634	-5.650	-5.665	-5.680	-5.695	-5.710	-5.724	-5.739	-5.753	-200
-190	-5.439	-5.456	-5.473	-5.489	-5.506	-5.522	-5.539	-5.555	-5.571	-5.587	-5.603	-190
-180	-5.261	-5.279	-5.297	-5.315	-5.333	-5.351	-5.369	-5.387	-5.404	-5.421	-5.439	-180
-170	-5.069	-5.089	-5.109	-5.128	-5.147	-5.167	-5.187	-5.205	-5.223	-5.242	-5.261	-170
-160	-4.865	-4.886	-4.907	-4.928	-4.948	-4.969	-4.989	-5.010	-5.030	-5.050	-5.069	-160
-150	-4.648	-4.670	-4.693	-4.715	-4.737	-4.758	-4.780	-4.801	-4.823	-4.844	-4.865	-150
-140	-4.419	-4.442	-4.466	-4.489	-4.512	-4.535	-4.558	-4.581	-4.603	-4.626	-4.648	-140
-130	-4.177	-4.202	-4.226	-4.251	-4.275	-4.299	-4.323	-4.347	-4.371	-4.395	-4.419	-130
-120	-3.923	-3.949	-3.974	-4.000	-4.026	-4.051	-4.077	-4.102	-4.127	-4.152	-4.177	-120
-110	-3.656	-3.684	-3.711	-3.737	-3.764	-3.791	-3.818	-3.844	-3.870	-3.897	-3.923	-110
-100	-3.378	-3.407	-3.435	-3.463	-3.491	-3.519	-3.547	-3.574	-3.602	-3.629	-3.656	-100

-90	-3.089	-3.118	-3.147	-3.177	-3.206	-3.235	-3.264	-3.293	-3.321	-3.350	-3.378	-90
-80	-2.788	-2.818	-2.849	-2.879	-2.909	-2.939	-2.970	-2.999	-3.029	-3.059	-3.089	-80
-70	-2.475	-2.507	-2.539	-2.570	-2.602	-2.633	-2.664	-2.695	-2.726	-2.757	-2.788	-70
-60	-2.152	-2.185	-2.218	-2.250	-2.283	-2.315	-2.348	-2.380	-2.412	-2.444	-2.475	-60
-50	-1.819	-1.853	-1.886	-1.920	-1.953	-1.987	-2.020	-2.053	-2.087	-2.120	-2.152	-50
-40	-1.475	-1.510	-1.544	-1.579	-1.614	-1.648	-1.682	-1.717	-1.751	-1.785	-1.819	-40
-30	-1.121	-1.157	-1.192	-1.228	-1.263	-1.299	-1.334	-1.370	-1.405	-1.440	-1.475	-30
-20	-0.757	-0.794	-0.830	-0.867	-0.903	-0.940	-0.976	-1.013	-1.049	-1.085	-1.121	-20
-10	-0.383	-0.421	-0.458	-0.496	-0.534	-0.571	-0.608	-0.646	-0.683	-0.720	-0.757	-10
0	0.000	-0.039	-0.077	-0.116	-0.154	-0.193	-0.231	-0.269	-0.307	-0.345	-0.383	0
0	0.000	0.039	0.078	0.117	0.156	0.195	0.234	0.273	0.312	0.351	0.391	0
10	0.391	0.430	0.470	0.510	0.549	0.589	0.629	0.669	0.709	0.749	0.789	10
20	0.789	0.830	0.870	0.911	0.951	0.992	1.032	1.073	1.114	1.155	1.196	20
30	1.196	1.237	1.279	1.320	1.361	1.403	1.444	1.486	1.528	1.569	1.611	30
40	1.611	1.653	1.695	1.738	1.780	1.822	1.865	1.907	1.950	1.992	2.035	40
50	2.035	2.078	2.121	2.164	2.207	2.250	2.294	2.337	2.380	2.424	2.467	50
60	2.467	2.511	2.555	2.599	2.643	2.687	2.731	2.775	2.819	2.864	2.908	60
70	2.908	2.953	2.997	3.042	3.087	3.131	3.176	3.221	3.226	3.312	3.357	70
80	3.357	3.402	3.447	3.493	3.538	3.584	3.630	3.676	3.721	3.767	3.813	80
90	3.813	3.859	3.906	3.952	3.998	4.044	4.091	4.137	4.184	4.231	4.277	90
100	4.277	4.324	4.371	4.418	4.465	4.512	4.559	4.607	4.654	4.701	4.749	100
110	4.749	4.796	4.844	4.891.	4.939	4.987	5.035	5.083	5.131	5.179	5.227	110
120	5.227	5.275	5.324	5.372	5.420	5.469	5.517	5.566	5.615	5.663	5.712	120
130	5.712	5.761	5.810	5.859	5.908	5.957	6.007	6.056	6.155	6.155	6.204	130

Table 6-3 *(Continued)*

°C	0	1	2	3	4	5	6	7	8	9	10	°C
					Thermoelectric Voltage in Absolute Mill Volts							
140	6.204	5.254	6.303	6.353	6.403	6.452	6.502	6.552	6.602	6.652	6.702	140
150	6.702	6.753	6.803	6.853	6.903	6.954	7.004	7.055	7.106	7.156	7.207	150
160	7.207	7.258	7.309	7.360	7.411	7.462	7.513	7.564	7.615	7.666	7.718	160
170	7.718	7.769	7.821	7.872	7.924	7.975	8.027	8.079	8.131	8.183	8.235	170
180	8.235	8.287	8.339	8.391	8.443	8.495	8.548	8.600	8.652	8.705	8.757	180
190	8.757	8.810	8.863	8.915	8.968	9.021	9.074	9.127	9.180	9.233	9.286	190
200	9.286	9.339	9.392	9.446	9.499	9.553	9.606	9.659	9.713	9.767	9.820	200
210	9.820	9.874	9.928	9.982	10.036	10.090	10.144	10.198	10.252	10.306	10.360	210
220	10.360	10.414	10.469	10.523	10.578	10.632	10.687	10.741	10.796	10.851	10.905	220
230	10.905	10.960	11.015	11.070	11.135	11.180	11.235	11.290	11.345	11.401	11.456	230
240	11.456	11.511	11.566	11.622	11.677	11.733	11.788	11.844	11.900	11.956	12.011	240
250	12.011	12.067	12.123	12.179	12.235	12.291	12.347	12.403	12.459	12.515	12.572	250
260	12.572	12.628	12.684	12.741	12.797	12.854	12.910	12.967	13.024	13.080	13.137	260
270	13.137	13.194	13.251	13.307	13.364	13.421	13.478	13.535	13.592	13.650	13.707	270
280	13.707	13.764	13.821	13.879	13.936	13.993	14.051	14.108	14.166	14.223	14.281	280
290	14.281	14.339	14.396	14.454	14.512	14.570	14.628	14.686	14.744	14.802	14.860	290
300	14.860	14.918	14.976	15.034	15.092	15.151	15.209	15.267	15.326	15.384	15.443	300
310	15.443	15.501	15.560	15.619	15.677	15.736	15.795	15.853	15.912	15.971	16.030	310
320	16.030	16.089	16.148	16.207	16.266	16.325	16.384	16.444	16.503	16.562	16.621	320
330	16.621	16.681	16.740	16.800	16.859	16.919	16.978	17.038	17.097	17.157	17.217	330

340	17.217	17.277	17.336	17.396	17.456	17.516	17.576	17.636	17.696	17.756	17.816	340
350	17.816	17.877	17.937	17.997	18.057	18.118	18.178	18.238	18.299	18.359	18.420	350
360	18.420	18.480	18.541	18.602	18.662	18.723	18.784	18.845	18.905	18.966	19.027	360
370	19.027	19.088	19.149	19.210	19.271	19.332	19.393	19.455	19.516	19.577	19.638	370
380	19.638	19.699	19.761	19.822	19.883	19.945	20.006	20.068	20.129	20.191	20.252	380
390	20.252	20.314	20.376	20.437	20.499	20.560	20.622	20.684	20.746	20.807	20.869	390
400	20.869											400

For a thermocouple with a Seebeck coefficient of α, the voltage output as a function of frequency is

$$dV = \alpha \Delta T = \frac{\alpha \varepsilon \phi_e}{K[1 + \omega^2 \tau_T^2]^{1/2}} \quad \text{(V)} \qquad (6.31)$$

The voltage responsivity (V W^{-1}) can now be determined.

$$\mathscr{R}_v = \frac{\alpha \varepsilon}{K[1 + \omega^2 \tau_T^2]^{1/2}} \quad \text{(V W}^{-1}) \qquad (6.32)$$

At very low frequencies, $\omega^2 \tau_T^2 \ll 1$, the voltage responsivity is

$$\mathscr{R}_v = \frac{\alpha \varepsilon}{K} \qquad (6.33)$$

Therefore, a small thermal conductance, a large Seebeck coefficient, and a large emissivity generate a large responsivity. A practical limitation to the performance of this detector is that it has a very slow response. Typically, chopping frequencies above 20 Hz cannot be used.

Although the responsivity is important, the more important parameter is the noise equivalent power (NEP). The noises associated with a thermocouple are Johnson noise and temperature noise, neither of which is a function of detector collecting area.

For the case of Johnson-noise-limited operation, recalling that NEP is the noise divided by the responsivity,

$$\text{NEP} = 4kTR \, \Delta f \sqrt{1 + \omega^2 \tau_T^2} \, \frac{K}{\alpha \varepsilon} \qquad (6.34)$$

Equation (6.34) defines what is needed for a low NEP. However, by minimizing K, the electrical resistance R is affected. The thermal conductance and the electrical conductance (σ) are related by

$$\frac{K}{\sigma T} = \text{constant} \qquad (6.35)$$

Therefore, the reduction in thermal conductance is balanced by the increase in electrical resistance (Willardson and Beer, 1970).

The spectral response is determined by the absorber and the window material, if present. Normally, the absorber is considered to be black and spectrally flat to 40 μm. For further discussion of black absorbers see Blevin and Geist (1974).

Since the resistance of thermocouples is so low, they are often coupled to a preamplifier through a transformer for impedance matching to obtain the lowest noise figure. The impedance is transformed by

$$\frac{Z_{\text{out}}}{Z_{\text{in}}} = \left(\frac{N_o}{N_{\text{in}}}\right)^2 \tag{6.36}$$

where Z_{in} = input impedance,

Z_{out} = output impedance,

N_o = number of wire turns on the output winding of the transformer,

N_{in} = number of wire turns on the input winding of the transformer.

6-5 THERMOPILES

Single thermocouples are not very practical as detectors. The responsivity can be increased when n thermocouples are placed in series; the responsivity is increased by n:

$$\mathcal{R}_v = \frac{n\alpha\varepsilon}{K} \tag{6.37}$$

Such a device is called a thermopile. There are two types of thermopiles: one made of a series of wires as suggested above and another is made by an evaporated thin-film approach.

The most commonly used materials for a thermopile are bismuth and antimony, which gives the best Seebeck coefficient for the stable metals. The thermopile is made of evaporated bismuth and antimony. The active junction area has deposited on it an energy-absorbing black paint. It is usually placed in a hermetically sealed container purged with argon or nitrogen gas.

BIBLIOGRAPHY

Blevin, W. R. and J. Geist, "Influence of Black Coatings on Pyroelectric Detectors," *Appl. Opt.* **13**(5), May (1974).

Cadoff, I. B. and E. Miller, *Thermoelectric Materials and Devices*, Reinhold, New York, 1960.

Dewaard, R. and E. M. Wormser, "Description and Properties of Various Thermal Detectors," *Proc. IRE* **47a**, 1508 (1959).

Egli, P. H., *Thermoelectricity*, Wiley, New York, 1960.

Kruse, P. W., L. D. McGlauchlin, and R. B. McQuistan, *Elements of Infrared Technology*, Wiley, New York, 1969.

Jones, R. Clark, "Noise in Radiation Detectors," *Proc. IRE* **47**(9), 1481 (1959).

Pollock, D. D., *The Theory and Properties of Thermocouples*, ASTM Special, Tech. Pub. 492, 1916 Race St., Philadelphia, PA 19103.

Putley, E. H., "The Ultimate Sensitivity of Sub-mm Detectors," *IR Physics* **4**(1), 1 (1964).

Smith, R. A., F. E. Jones, and R. F. Chasmer, *Detection and Measurement of Infrared Radiation*, Clarendon Press, Oxford, 1957, pp. 213–215.

Travers, D., "Thermocouple Chip Includes Ice-point Reference," *Electronic Products*, **25**(2), 80, June (1982).

Wasa, K., T. Tohda, Y. Kasahara, and S. Kayakawa, "Highly Reliable Temperature Sensor Using rf-Sputtered SiC Thin Film," *Rev. Sci. Instrum.* **50**(9), 1084–1088, Sept. (1979).

Willardson, R. K., and A. C. Beer, *Semiconductors and Semimetals*, Vol. 5, Academic Press, New York, 1970.

Wormser, E. M., "Properties of Thermistor Infrared Detectors," *J. Opt. Soc. Am.* **43**, 15 (1953).

Yariv, A., *Introduction to Optical Electronics*, 2nd ed., Holt, Reinhart, and Winston, New York, 1976.

PROBLEMS

6-1. Discuss the thermopile as a thermoelectric cooler.

6-2. Radiation thermopiles are frequently evacuated. Why?

6-3. A thermal detector is at and surrounded by a temperature of 300 K. The radiation-limited NEP is 10^{-8} W. If the detector temperature was reduced to 150 K, what would be the new NEP value?

6-4. Derive the relationship between thermal time constant and responsivity for a thermal detector.

6-5. Derive the NEP expression for a thermal detector if the dominant noise is due to temperature fluctuations.

6-6. List the factors that contribute to raising a cooled detector's temperature.

6-7. Show how the performance can be improved by immersion of a detector into a hemispherical lens. Assume a germanium lens ($n = 4$). Calculate the gain in signal-to-noise output with the lens:

    a. Internally noise-limited detector.

    b. BLIP-limited condition.

6-8. For a BLIP thermal detector at 4 K, what is the best NEP possible with 77 K background? $A_d = 1\text{mm}^2$, $\Delta f = 1$ Hz.

CHAPTER 7

BOLOMETERS

The bolometer has been in use since 1880 when it was invented by Langley (1881) to detect both visible and infrared radiation. Although other thermal devices have been developed since that time, the bolometer remains one of the most used infrared detectors. The bolometer need not require cooling, it has a flat spectral response limited only by the filter being used, it can be made to withstand rugged conditions, and cooled bolometers can approach photon-limited performance. The primary attraction, however, is the ease with which it may be operated.

The principle of operating for a bolometer is that a temperature change produced by the absorption of radiation causes a change in electrical resistance of the material used to fabricate the bolometer. This change in resistance can be used to sense radiation just as in the photoconductor, however, the basic detection mechanisms are different. In the case of a bolometer, radiant power produces heat within the material, which in turn produces the resistance change. There is not a direct photon–lattice interaction.

The resistance change of a material is specified in terms of a temperature change of 1° C. Mathematically,

$$\alpha = \frac{1}{R}\frac{dR}{dT} \tag{7.1}$$

where α = temperature coefficient of resistance,
R = resistance,
T = temperature.

Bolometers may be divided into five types. The most commonly used

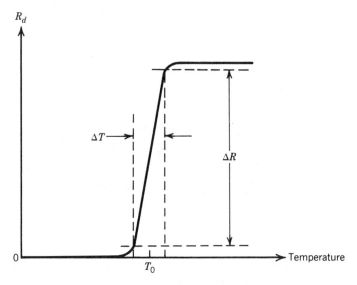

Figure 7-1 Superconducting transition edge.

are the metal, the thermistor, and the semiconductor bolometers. The composite bolometer has only recently been developed and looks promising for low-noise, low-temperature infrared detection.

A fifth type is the superconducting bolometer. This bolometer operates on a conductivity transition in which the resistance changes dramatically (orders of magnitude) over a transition temperature range (refer to Fig. 7-1).

If the temperature of the bolometer is held near the midpoint of this transition (typically 1.3 K), then radiation can produce sufficient heat to cause large resistance changes. Because of the stringent temperature control that is required, this device has not been used extensively as a detector.

7-1 METAL BOLOMETER

The metal bolometer is the first kind of bolometer ever used (Langley, 1881). A picture of this early bolometer is in Fig. 7-2. Typical materials are nickel, bismuth, or platinum. Being made of metal, these bolometers need to be small so that the heat capacity is low enough to allow reasonable sensitivity. Most metal bolometers are formed as film strips, about 100–500 Å thick, via vacuum evaporation or sputtering. Since such

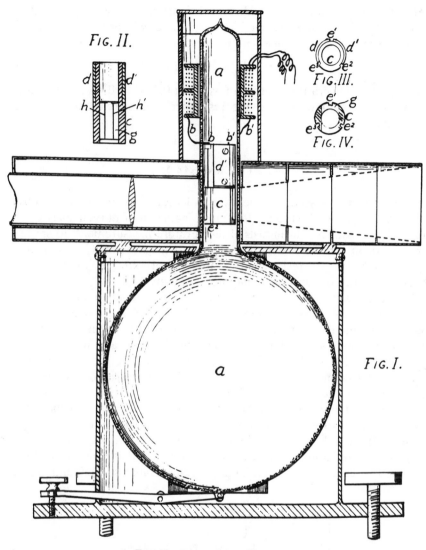

FIG. II.

FIG. III.

FIG. IV.

FIG. I.

1. THE MAKE-UP OF THE BOLOMETER

Sensitive strips *h, h'* mounted on the copper blocks *c, d'* form two arms of a Wheatstone's bridge, completed by the two coils *b, b'*. Solar spectral rays are admitted by a vestibule with diaphragms, and may be adjusted by the eyepiece. The vessel, *a*, is highly evacuated.

Figure 7-2 Very early bolometer. (By permission of Smithsonian Institution Press from *Annals of the Astrophysical Observatory of the Smithsonian Institution*, vol. 6, Washington, D.C., 1942).

153

metal strips highly reflect the radiation they are detecting, they are often coated with a black absorber such as evaporated gold or platinum black.

The variation of resistance with temperature is expressed as

$$R = R_0 [1 + \alpha(T - T_0)] \tag{7.2}$$

where α, the temperature coefficient, is about 0.5%/°C. This temperature coefficient is positive, and so the resistance will increase for rises in temperature.

7-2   THERMISTOR

The thermistor is a second-generation bolometer and perhaps the most popular. It has found wide application, ranging from burglar alarms, to fire-detection systems, to industrial temperature measurement, to space-borne horizon sensors and radiometers. When hermetically sealed, this type of detector can resist such environmental extremes as vibration, shock, temperature variations, and high humidity.

The temperature-sensitive element in a thermistor bolometer is typically made of wafers of manganese, cobalt, and nickel oxides sintered together and mounted on an electrically insulating but thermally conducting material such as sapphire. The substrate serves as a heat sink. When incident radiation is absorbed, the temperature of the metal oxide film increases. This temperature increase raises the free carrier concentration, thus decreasing the electrical resistance. A thermistor is often fabricated so as to have a temperature coefficient as high as 5%/°C. This coefficient is not constant but varies as $1/T^2$ (Shive, 1959). The usual construction uses a matched pair of devices for a single unit. One of the pair of thermistors is shielded from radiation and fitted into a bridge such that it acts as the load resistor. (Refer to Fig. 7-3.) This arrangement allows maximum signal

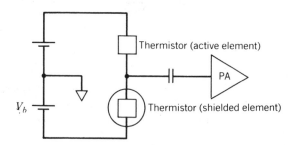

Figure 7-3   Biasing configuration for thermistor bolometer.

transfer over ambient temperature changes. The result can be a dynamic range of a million to one.

The responsivity of a thermistor is a function of chip size, bias, and time constant; as an example, a 1 mm² detector with a 1 msec time constant can be expected to have a responsivity of about 450 V W⁻¹.

The time constant varies around 1–5 msec. Being a thermal detector, greater thermal conductivity ensures faster response but simultaneously prevents the film from reaching higher temperatures, lowering the responsivity. Thus, the classical trade-off between sensitivity and frequency response exists. For an NEP = 10^{-10} W, the time constant may be about 100 msec, whereas for an NEP = 10^{-8} W, the time constant would be 1 msec.

Table 7-1 gives some typical values of parameters for a thermistor operating at room temperature.

Table 7-1

Size	1–3 mm²
Resistance[a]	250 kΩ–2.5 MΩ
Time constant	1–100 msec
NEP	10^{-10}–10^{-8} W
Spectral response[b]	0.4–10 μm

[a]This is for square elements, the resistance of rectangular flakes is given by length over width (l/w) times the square resistance.
[b]Depends on window transmission and emissivity of absorber.

The noise sources associated with a thermistor are Johnson noise, thermal fluctuation noise, $1/f$ noise, and preamplifier noise. The thermal fluctuation noise is relatively low and not important in real systems. The preamplifier noise is due to the power supply bias and can be reduced by using good noise-free batteries. The Johnson noise is usually the limiting noise of operation.

Since the thermistor is Johnson noise limited, an improvement in sensitivity (NEP) can be realized by placing a hemispherical or hyperspherical lens over its surface (Jones, 1962). This procedure would not improve the signal-to-noise ratio if the detector is photon noise limited. The detector must be optically coupled with the lens so that total internal

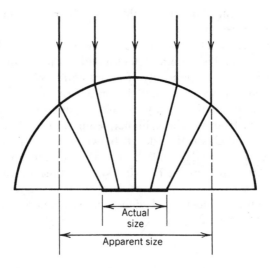

Figure 7-4   Immersion lens on thermistor.

reflection and Fresnel losses will not occur. This can best be accomplished by depositing the detector directly on the plane side of the lens
(Astheimer, 1983).

As shown in Fig. 7-4 the ray directed toward the center of the
thermistor is not refracted (bent) at the lens interface; however, the rays
directed to the edge of the detector are refracted by the lens, giving the
detector an appearance of being larger by a factor of n, where n is the
index of refraction of the lens. Since the detector is a two-dimensional
device, the virtual area increase is n^2. As a result, the signal energy
collected on the thermistor is increased by n^2 as is the signal-to-noise
ratio. The requirements on the lens material are that it have as large an
index of refraction as possible, and it should be electrically insulating so as
not to short the thermistor film. Germanium, silicon, and arsenic tri-
selenide are among the most useful thermistor materials.

7-3 SEMICONDUCTOR BOLOMETERS

In the normal operation of a bolometer, an accurately controlled bias
current I is passed through the element. Thus, for a radiation input
causing a change dR_d in the bolometer resistance, there will be an output
voltage

$$\Delta V = I(dR_d) = I\left(\frac{1}{R_d}\frac{dR_d}{dT}\right) R_d \, dT$$

$$= I\alpha R_d \Delta T \tag{7.3}$$

From the equation for ΔT [Eq. (6.8)] this becomes

$$\Delta V = \frac{2I\Delta\phi_e\varepsilon\alpha R_d}{K(1 + \omega^2 H^2/K^2)} \tag{7.4}$$

and the voltage responsivity is

$$\mathscr{R}_V = \frac{2I\varepsilon\alpha R_d}{K(1 + \omega^2 H^2/K^2)} \tag{7.5}$$

One sees that a high responsivity bolometer requires

a high value for the temperature coefficient of resistance α

a high value of bolometer resistance R_d that is still compatible with a low-noise preamplifier

a high absorptance

a low thermal capacitance H

At very low temperatures it is possible to obtain both a much larger value for the temperature coefficient of resistance as well as a much smaller specific heat. If the detector and its surroundings are cooled to a very low temperature, the ultimate sensitivity can be orders higher than for a room-temperature bolometer. Very-high-performance low-temperature semiconducting bolometers have been developed. Semiconductors, when lightly doped with suitable impurities, are an excellent material since they will respond to just beyond 100 μm, and they are good absorbers. By comparison, photon detectors only give their optimum performance once a relatively narrow spectral band and, at the time of development of low-temperature bolometers, the longest spectral response of a photoconductive detector was barely 5 μm.

Low (1961) was the first to develop the Ge bolometer operating in the liquid-helium range. By using single-crystal gallium-doped germanium, he obtained sensitivities close to the theoretical limits. For small apertures at 4.2 K the inherent detector noise is divided equally between Johnson noise and phonon noise (i.e., temperature fluctuations resulting from fluctuations in the incident radiation; this is a photon-induced noise). At larger apertures the photon noise may predominate over the inherent detector noise, depending on the temperature of the background. For example, a

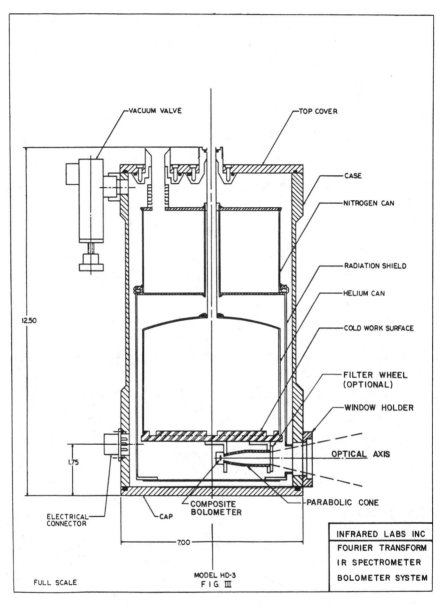

Figure 7-5 Dual cryogenic Dewar (courtesy Infrared Labs, Inc.).

detector with field of view of 180° and a background temperature of 300 K, at 4.2 K operation, is effectively background limited. The background limit can be achieved for even smaller apertures and lower background temperatures if the detector is cooled below 4.2 K.

The low temperature bolometer is housed in a dual cryogenic dewar as shown in Fig. 7-5. The bolometer is held at a temperature of liquid helium or pumped helium (<4.2 K). The bolometer elements are cut from suitably doped single-crystal germanium, and are lapped and etched to the desired thickness. Because of the piezoresistance of germanium at low temperature, it is necessary to mount the element in a strain-free manner. The element is supported in its evacuated housing by the two electrical leads. These leads provide the only appreciable thermal contact with the bath. Figure 7-6 shows the element construction. This method of construction allows the thermal conductance to be varied several decades by changing the diameter, length, and composition of the leads. The thermal conductivity of the germanium is sufficiently high so that the temperature at the center of the element remains very nearly equal to that at the ends. The load resistor is placed in the cryostat to reduce Johnson noise. The surface of the germanium element is blackened to get approximately unit emissivity over a wavelength range of 0.4–40 μm.

Consider the heat-equilibrium equation, which determines the temperature of the element:

$$H \frac{dT}{dt} + K_0(T_d - T_0) = I^2 R_d$$

This is the heat-balance equation when the thermal conduction of the

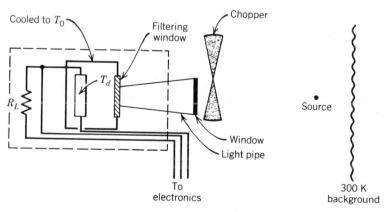

Figure 7-6 Bolometer element suspended on wires.

leads just equals the current heating effects ($I^2 R_d$). Radiation incident on the detector causes additional heating. The bolometer is at a temperature T, which is assumed to be the same at all points of the element. Then we have

$$H\frac{dT}{dt} + K(T - T_0) = P + \phi_e^b + \varepsilon\Delta\phi_e \qquad (7.6)$$

where H = thermal capacity of the bolometer element (J K^{-1}) (i.e., the specific heat times the mass: $C_p m$),
K = average thermal conductance from the element to heat sink (W K^{-1}),
ϕ_e^b = background radiant energy,
$\Delta\phi_e$ = signal radiant energy,
P = thermal power generated by bolometer self-heating ($I^2 R_d$),
T_0 = heat sink temperature (ambient),
T = bolometer temperature,
ε = detector emissivity.

Equation (7.6) accurately describes bolometer operation only if $\Delta\phi_e$ is much less than the electrical (VI) power dissipation ($\Delta\phi_e \ll P$) and the background radiation power is less than or equal to the electrical power dissipation ($\phi_c^b \le P$). Redefining an effective bath temperature T_0' so that $K(T_0' - T_0) = \phi^b$, Eq. (7.6) may be written as

$$H\frac{dT}{dt} + K(T - T_0') = P + \Delta\phi \qquad (7.7)$$

We see that a higher effective bath temperature results in degradation of performance. If $\phi^b \ll P$, we have $T_0' \sim T_0$ and small-signal analysis can be applied.

In the steady state, $\Delta\phi = 0$, the time-independent temperature T of the bolometer is found from

$$K(T - T_0') = I^2 R \qquad (7.8)$$

When a nonzero signal $\Delta\phi$ is incident on the bolometer, the temperature increases to $T + \Delta T$ and the Joule heating power will change by $\Delta P = (dP/dt)\Delta T$. Therefore, Eq. (7.7) becomes

$$H\frac{dT}{dt} + K\Delta T = \frac{dP}{dT}\Delta T + \Delta\phi \qquad (7.9)$$

The rate of change of Joule heating with temperature, dP/dT, depends on the biasing circuit. A typical configuration is shown in Fig. 7-7. For a

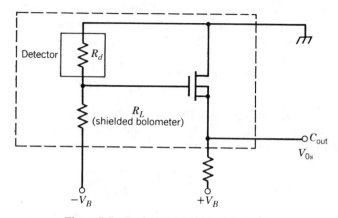

Figure 7-7   Typical biasing of bolometer.

review of preamplifiers used with low-temperature bolometers see Geobel (1977).

Rearranging the rate of change of power with temperature:

$$\frac{dP}{dT}\Delta T = \frac{d(I^2 R)}{dT} = V_B^2 \frac{d}{dT}\left[\frac{R_d}{(R_L + R_d)^2}\right]\Delta T$$

$$= I^2\left(\frac{dR_d}{dT}\right)\frac{(R_L - R_d)}{(R_L + R_d)}\Delta T$$

$$= \frac{\alpha P(R_L - R_d)}{(R_L + R_d)}\Delta T$$

Using Eq. (7.8), then,

$$\frac{dP}{dT}\Delta T = \alpha G(T - T_0)\frac{(R_L - R_d)}{(R_L + R_d)} \tag{7.10}$$

where α is the temperature coefficient of resistance defined by Eq. (7.1). Therefore, we may rewrite Eq. (7.9) in the form

$$H\frac{d\Delta T}{dT} + K_e\Delta T = \Delta\phi \tag{7.11}$$

where K_e is the effective thermal conductance and is given by

$$K_e = K - \alpha K(T - T_0)\frac{(R_L - R_d)}{(R_L + R_d)} \tag{7.12}$$

Assuming a periodic form for the radiant power, similar to the approach used in Chapter 6, allows one to find the following solution for the temperature change ΔT from Eq. (7.11):

$$\Delta T = \Delta T_0 e^{-K_e t/H} + \frac{2\varepsilon\phi_{e0}}{[K_e^2 + \omega^2 H^2]^{1/2}} \qquad (7.13)$$

where a thermal time constant can be defined as K_e/H.

The first term of Eq. (7.13) exponentially decays as K_e increases from an initial positive value. However, if K_e is negative, that is,

$$K < K_0(T_d - T_0) \propto \frac{R_L - R_d}{R_d + R_L} \qquad (7.14)$$

then as K_e decreases, the first term of Eq. (7.13) increases exponentially, and burnout of the bolometer may occur. Bolometer burnout can occur when the heat added to the bolometer exceeds the heat conducted away by the electrical lead wires.

The second term of Eq. (7.13) dominates during normal operation of the bolometer:

$$\Delta T \approx \frac{2\varepsilon\phi_{e0}}{[K_e^2 + \omega^2 H^2]^{1/2}} \qquad (7.15)$$

where $K_e \gg H$. From Eq. (7.1), the fractional change in bolometer resistance caused by a temperature change is

$$\frac{dR_d}{R_d} = \alpha \, dT. \qquad (7.16)$$

Combining Eqs. (7.15) and (7.16),

$$\frac{dR_d}{R_d} = \frac{2\alpha\varepsilon\phi_{e0}}{[K_e^2 + \omega^2 H^2]^{1/2}} \qquad (7.17)$$

From the biasing configuration of Fig. 7-7, the voltage signal out is

$$dV = \frac{V_B R_L}{(R_L + R_d)^2} \qquad (7.18)$$

Setting $R_L = R_d$ and substituting for dR_d,

$$dV = \frac{V_B \alpha\varepsilon\phi_{e0}}{2[K_e^2 + \omega^2 H^2]^{1/2}} \qquad (7.19)$$

Therefore the voltage responsivity is

$$\mathcal{R}_V = \frac{dV}{\phi_e} = \frac{\alpha \varepsilon V_B}{2[K_e^2 + \omega^2 H^2]^{1/2}} \tag{7.20}$$

7-4 LOW TEMPERATURE GERMANIUM BOLOMETER

The germanium bolometer has been the most commonly used semiconductor bolometer. When operated at cryogenic temperatures, performance near the theoretical limit can be achieved over the wavelength range from 5 to 100 μm. The germanium bolometer is therefore an extremely versatile laboratory device when only one high-performance infrared detector is desired. In the far-infrared range from 15 to 100 μm, semiconductor bolometers in general, and germanium in particular, are the detectors of choice. The use of these far-infrared wavelengths is largely restricted to astronomy performed at high altitude or from a space platform owing to the high level of atmospheric absorption.

An empirical relation for the resistance versus temperature of the bolometric material has been deduced from measurements between 1.1 K and 4.2 K,

$$R(T) = R_0 \left(\frac{T_0}{T}\right)^A \tag{7.21}$$

where R_0 is the resistance at temperature T_0. The constant A is approximately equal to 4 but must be determined for each bolometer. The temperature coefficient of resistance α can thus be written as

$$\alpha(T) = \frac{1}{R}\left(\frac{dR}{dT}\right) = \frac{-A}{T} \tag{7.22}$$

Equation (7.22) shows that α increases as T is reduced. Since the responsivity is proportional to α and it is desirable to make the responsivity as large as possible, operation below 4 K is usually advantageous. There are other advantages: the specific heat falls as the temperature is reduced, and lowering the temperature will increase the resistance R. If R is not too large, this may also be advantageous, but if R rises to such a large value that matching to the amplifier becomes difficult, there may be an optimum value for the operating temperature.

It can be shown (Low, 1961) that the responsivity of Eq. (7.20) can be

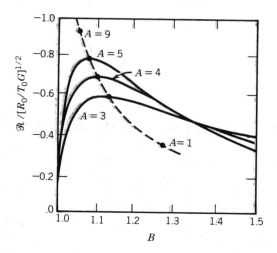

Figure 7-8 Equation (7.23) is plotted for three values of A. The dashed line shows the position of maxima for other values of A (after Low, 1963).

rewritten:

$$\mathscr{R}_V(B) = -\left(\frac{A(B-1)}{(A+1)B - A^2 B^A}\right)^{1/2}\left(\frac{R_0}{T_0 G}\right)^{1/2} \qquad (7.23)$$

where $B = T/T_0$. For a given bath temperature T_0, $\mathscr{R}_V$ has a maximum value that depends only on the constant A. In Fig. 7-8 Eq. (7.23) is plotted for three values of A; the dashed line shows the position of the maximum for other values of A. For $A = 4$, the optimum value of B is 1.1, giving

$$\mathscr{R}_{max} = -0.7\left(\frac{R_0}{T_0 K}\right)^{1/2} \qquad (7.24)$$

and

$$\tau = \frac{\tau' B}{B + A(B-1)} = 0.73\tau' \qquad (7.25)$$

From Eq. (7.8), the optimum value of P is $0.1 T_0 K$. The Johnson noise and the phonon noise can be computed by using this approximate expression for $\mathscr{R}_{max}$. The equivalent Johnson noise power

$$\text{NEP} = \frac{(4kT_0 R_0 \Delta f)^{1/2}}{\mathscr{R}_{max}} \simeq \frac{(4kT_0^2 K \Delta f)^{1/2}}{0.7} \qquad (7.26)$$

where k = Boltzmann constant,
Δf = equivalent noise bandwidth.

$$\text{The photon noise power} = (16kT_0^2K\,\Delta f)^{1/2} \qquad (7.27)$$

The NEP, in the absence of photon noise,

$$\text{NEP} \simeq 2.86\,T_0\sqrt{kK\Delta f} \qquad (7.28)$$

Figure 7-9 shows some NEP values versus thermal conductance (K). Thus the inherent NEP depends only on the bath temperature T_0 and the value of K. Since K is the thermal conductance of the leads, the NEP is independent of the dimensions of the detector element. This has the important result that, unlike most detectors, the NEP does not vary as the square root of the area. Therefore, the specific detectivity D^* cannot be used as a valid means of comparison with other detectors. But if the thickness is fixed, the product of NEP and $\sqrt{\tau}$ does vary as the square root of the area.

The characteristics of a typical low-temperature germanium bolometer are summarized in Table 7-2. A value for D^* is included, but care should be taken in using this parameter.

Table 7-2

Characteristics of a Typical Low-Temperature Ge Bolometer

T_0 (K)	2.15
A_d (cm²)	0.15
t (cm)	0.012
R_L (Ω)	5.0×10^4
R_d (Ω)	1.2×10^4
$\mathscr{R}_V$ (V W^{-1})	4.5×10^3
τ (μsec)	400
f (Hz)	200
K (μW K^{-1})	183
NEP (W)	5×10^{-13}
D^*	8×10^{11}

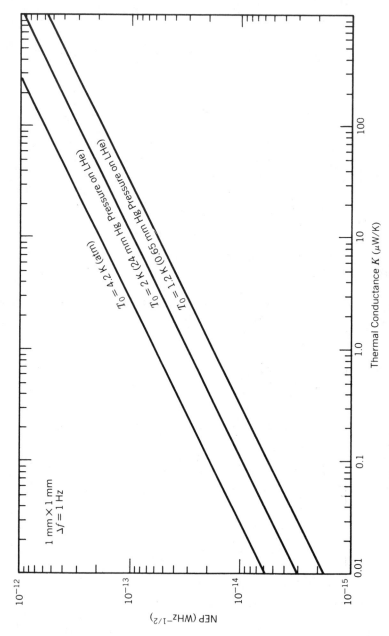

Figure 7-9 NEP versus thermal conductance (K) of a Ge bolometer.

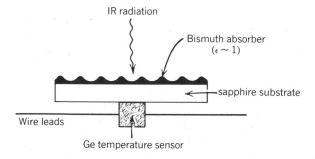

Figure 7-10 Composite bolometer.

7-5 COMPOSITE BOLOMETER

The composite bolometer came into being because a lower heat capacity (C) was needed to reduce the time constant (Clarke, Richards, and Yeh, 1977). It turns out that sapphire has about 1/60th the heat capacity of germanium. This means that a larger detector area can be made with no corresponding frequency response loss. Diamond has a heat capacity of approximately 1/600th that of germanium, thus much larger active areas can be achieved.

The composite bolometer is made up of three parts, the radiation-absorbing material; the substrate, which determines its active area, and the temperature sensor, as shown in Fig. 7-10. The absorber is made of a thin film whose thickness and composition are adjusted so that the emissivity is very high into the hundreds of microns wavelength region. Black bismuth and nichrome absorbers have been used. The temperature sensor shown as germanium in Fig. 7-10 is bonded both mechanically and thermally to the substrate via an epoxy or varnish. Thus the substrate and film combination act as an efficient absorbing element of large effective area and low heat capacity for a temperature sensor that can be made very small.

7-6 INTERFACING THERMAL DETECTORS

Typical electrical operation of a thermal detector was shown in Fig. 7-7. This biasing circuit is similar to that used with a photoconductor. The resistance varies with dT, as was derived in Chapter 6. Therefore,

$$dR_d = \frac{R_d \, \phi_{e0}}{K[1 + \omega^2 H^2 / K^2]^{1/2}} \tag{7.29}$$

From our photoconductor evaluation [Eq. (4.16)] of the above circuit

$$dV_0 = \frac{V_B R_L}{(R_L + R_d)^2} dR_d \qquad (7.30)$$

Substituting for dR_d

$$dV_0 = \frac{V_B R_L}{(R_L + R_d)^2} \frac{R_d \phi_{e0}}{K(1 + \omega^2 H^2/K^2)^{1/2}} \qquad (7.31)$$

For maximum responsivity we desire $R_L = R_d$. If two detectors are made identically, one can be used as a detector and the other as the load resistor by shielding it from optical radiation. The responsivity is then

$$\mathscr{R}_V = \frac{V_B \alpha}{4K(1 + \omega^2 H^2/K^2)^{1/2}} \qquad (7.32)$$

where $\omega = 2\pi f$,
$\tau = H/K$ (thermal time constant).

The negative sign cancels with negative α:

$$\alpha = \frac{\beta}{T^2}$$

$$\mathscr{R}_V = \frac{V_B \alpha}{4K(1 + \omega^2 \tau^2)^{1/2}}$$

Taking the Fourier transform to express Eq. (7.32) in the time domain,

$$\mathscr{R}_V = \frac{V_B \alpha}{4K}[1 - e^{-t/\tau}] \qquad (7.33)$$

An important conclusion which one can make from this equation is that one cannot have a high-responsivity fast thermal detector.

BIBLIOGRAPHY

Astheimer, R. W., "Thermistor Infrared Detectors," 27th SPIE International Symposium, Vol. 443.

Clark, Richards, and Yeh, "Composite Superconducting Transition Edge Bolometer," Appl. Phys. Lett. 30(12), June (1977).

Johnson, C., A. W. Davidson, and F. J. Low, "Germanium and Germanium-Diamond Bolometers," Optical Engineering, 19(2), March/April (1980).

Geobel, J. H., "Liquid Helium Cooled Preamplifier," Rev. Sci. Instrum. 48(4), April (1977).

Langley, S. P., Nature, 25(14) (1881).

Low, F. J., "Low-Temperature Germanium Bolometer," *J. Opt. Soc. Am.* **51**(14), November (1961).

Low, F. J. and A. R. Hoffman, "The Detectivity of Cryogenic Bolometers," *Applied Optics* **2**(6), June (1963).

Nishioka, Richards, and Woody, "Composite Bolometers for Submillimeter Wavelengths," *Applied Optics* **17**(10), May (1978).

Shive, J. M., *Semiconductor Detectors*, Van Nostrand, New York, 1959.

Wolfe, W. L., and G. Zissis, *The Infrared Handbook*, Office of Naval Research, Washington, D.C.

PROBLEMS

7-1.   Explain why $1/f$ noise could be a bigger problem for thermal detectors than photoconductors.

7-2.   What is the exact change in D^*_{BLIP} if an immersion lens is used over a detector which is background noise limited?

7-3.   Explain burnout of a bolometer in terms of negative resistance.

7-4.   Compare and contrast a bolometer and a photoconductor.

7-5.   Discuss the trade-offs of placing a detector at a field stop or exit pupil (image of aperture stop).

7-6.   How would you prove the existence of an infrared rainbow? How far do you think its spectrum would go into the infrared?

7-7.   Why would you want to chop an infrared source?

7-8.   If a germanium bolometer operated at 2 K has a load curve that can be approximated by $I(\mu A) = 0.5 E_B^2$, what is its predicted responsivity and NEP for optimum bias of 2 V ($E_B = 2$ V)? What is the projected NEP for a temperature of 0.5 K?

7-9.   It is normally assumed that the thermal conductivity is about 10 times the incident radiation from the background

$$GT_0 = 10\phi_e^B$$

For the two temperatures ($T_0 = 2.0$ and 0.5), what background photon flux (ϕ_p^B) does this correspond to (detector is sensitive to all λ, background fills 2π sr, average energy of a blackbody photon is $2.75kT$)? $G = 2 \times 10^{-6}\,\mathrm{WK}^{-1}$.

7-10. What is the significance of the resistance-area (RA) product for a bolometer?

7-11. Why do photodetectors seem to require cooling for long-wavelength infrared-radiation detection (40 μm) while thermistors do not?

7-12. Derive the NEP expression for the low-temperature bolometer.

7-13.   For a bolometer that is connected as shown below the load resistor is a shielded or covered bolometer. Assume that different resistance bolometers are used as the detector and load resistors with values as shown below:

R_B	R_L
10	90
50	50
90	10

In each case, the temperature coefficient α (relative change of resistance with temperature)

$$\alpha = \frac{1}{R}\frac{\partial R}{\partial T}$$

is the same. Compare the relative voltage responsivity ($\mathscr{R}_V$) for these three cases. Which of the above conditions are optimum and why?

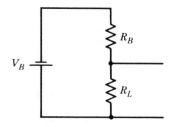

CHAPTER 8

PYROELECTRIC DETECTORS

The growing requirement for sensitive infrared detectors that operate at room temperature motivated the development of the pyroelectric detector (Cooper, 1962). the pyroelectric detector is important because it has proven to outperform other uncooled thermal detectors (Roundy and Byer, 1972). The most important pyroelectric detectors consist of materials of triglycerine sulfite (TGS), strontium barium niobate, polyvinylidene fluoride, and lithium tantalate. The most recent work on deutrated TGS and α-alinine TGS has been to increase the Curie temperature and responsivity.

8-1 BACKGROUND

The basis of operation of the pyroelectric material is that it possesses a spontaneous or permanent polarization P_s, in the absence of an applied electric field. This spontaneous polarization is governed by the material temperature and by the crystal symmetry and bonding. The known pyroelectric materials are dielectrics, and the change in polarization results in a proportional change in dielectric constant. This temperature-dependent spontaneous polarization causes charge to flow if connected to an external circuit. Typically, an electric field is applied to the detector causing the polarization to align along a preferred axis of the material. This process is called "poling" of the device. When radiation on the detector causes a heating, thus an expansion of the crystal lattice spacing, a change in electrical polarization occurs. This produces a change in the charge on the electrodes, and charge flow will occur in an external circuit.

Table 8-1

Pyroelectric Materials

Material	Pyroelectric Coefficient, P [C cm^{-2} K^{-1}]	Dielectric Constant, ε	Specific Heat, C [J g^{-1} K^{-1}]	Thermal Conductivity, K [W cm^{-1} K^{-1}]	Density, ρ [g cm^{-3}]
Turmaline	4×10^{-10}				
BaTiO$_3$	2×10^{-8}	160 ($\parallel$ polar axis) 4100 ($\perp$ polar axis)	0.5	9×10^{-3}	6.0
TGS	$(2\text{–}3.5) \times 10^{-8}$	25–50	0.97	6.8×10^{-3}	1.69
Li$_2$SO$_4$H$_2$O	1.0×10^{-8}	10	~0.4	17×10^{-3}	2.05
LiNbO$_3$	4×10^{-9}	30 ($\parallel$) 75 ($\perp$)			4.64
LiTaO$_3$	6×10^{-9}	58			
SbSI	2.6×10^{-7}	10^4	0.29		8.2
NaNO$_3$	1.2×10^{-8}	8.0	0.96		2.1

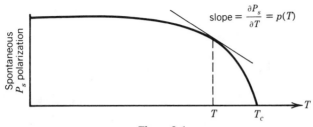

Figure 8-1

The pyroelectric coefficient p is a measure of the rate of change of electric polarization with respect to temperature change, or

$$p = \frac{\partial P_s}{\partial T} \tag{8.1}$$

Table 8-1 gives some relevant parameters of materials that exhibit the pyroelectric effect at 300 K.

In the static case of constant temperature, the material aligns itself internally to display no net charge difference. Therefore, the pyroelectric detector is a derivative or change-responding detector, with current

$$I = p(T)A_d \frac{\partial T}{\partial t} \tag{8.2}$$

where $p(T)$ = pyroelectric coefficient,
$\quad A_d$ = detector area,
$\quad \partial T/\partial t$ = temperature rate of change.

The pyroelectric coefficient is proportional to the slope of the polarization curve as shown in Fig. 8-1. Since the polarization charge is a function of temperature, so is the pyroelectric coefficient. The pyroelectric coefficient is plotted versus temperature in Fig. 8-2. Obviously, one wants the largest value to get maximum response.

The phase transition due to heating occurs when the material is no longer polarized at the Curie temperature (T_C) as indicated in Fig. 8-2. In order to make the material operate as a detector again once it has lost its polarization, a "poling" process is required. Possible causes of loss of polarization other than raising the temperature above the Curie temperature are age or mechanical shock of the device. The poling process consists of heating the detector material above the Curie temperature, applying a large bias to it (e.g., 1000 V cm^{-1}) and then slowly cooling the detector down to room temperature with the bias applied.

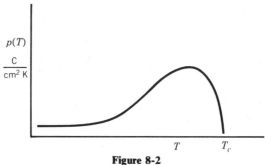

Figure 8-2

8-2   RESPONSIVITY OF PYROELECTRIC DETECTORS

The equivalent circuit of the pyroelectric detector is shown in Fig. 8-3. The polarization charge produced by a temperature change causes a current that is a function of frequency

$$i_d = \omega p A_d \Delta T \tag{8.3}$$

where $\omega = 2\pi f$ = radian frequency,
   p = pyroelectric coefficient,
   A_d = detector area,
   ΔT = temperature change.

Although a current responsivity can be calculated, we will direct our attention to the voltage responsivity. The output voltage signal (V_0) is the current (i_d) multiplied by the impedance:

$$V_0 = i_d \frac{R_d(1/j\omega C_d)}{R_d + 1/j\omega C_d} = \frac{i_d R_d}{\sqrt{1 + \omega^2 R_d^2 C_d^2}} \tag{8.4}$$

Substituting for i_d,

$$V_0 = \frac{\omega p A_d R_d (\Delta T)}{\sqrt{1 + \omega^2 R_d^2 C_d^2}} \tag{8.5}$$

where $R_d C_d$ is the electrical time constant

$$\tau_E = R_d C_d \tag{8.6}$$

Now an expression for the temperature change must be found.
   Following the method of Chapter 6, the heat-balance equation is used

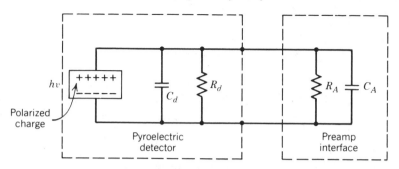

Figure 8-3   Pyroelectric equivalent circuit.

to find the expression for ΔT:

$$\text{heat absorbed} = [\text{heat to warm detector}] - [\text{heat losses}] \qquad (8.7)$$

The differential equation for heat flow in one dimension is

$$\phi_e(t) = C_T \frac{d(\Delta T)}{dt} + \frac{\Delta T}{R_T} + \sigma(T_d^4 - T_a^4) \qquad (8.8)$$

where R_T = thermal resistance to heat sink,
$\quad C_T$ = heat capacity of detector,
$\quad T_d$ = detector temperature,
$\quad T_a$ = ambient temperature,
$\quad \phi_e(t)$ = incident radiant power as a function of time.

The incident radiant power is

$$\phi_e(t) = \frac{\phi_e}{2}(e^{j\omega t} + 1) \qquad (8.9)$$

As shown in Chapter 6, the solution for ΔT is (neglecting radiation loss)

$$\Delta T = \frac{\varepsilon \phi_e R_T}{\sqrt{1 + \omega^2 R_T^2 C_T^2}} \qquad (8.10)$$

where the thermal time constant (τ_T) can be expressed as

$$\tau_T = R_T C_T \qquad (8.11)$$

Substituting for ΔT in the signal voltage expression Eq. (8.5):

$$V_0 = \frac{\omega p A_d R_d}{\sqrt{1 + \omega^2 \tau_E^2}} \frac{\varepsilon \phi_e R_T}{\sqrt{1 + \omega^2 \tau_T^2}} \qquad (8.12)$$

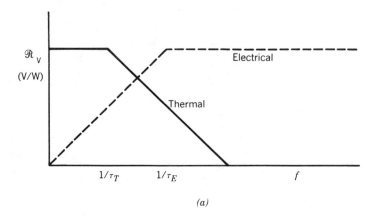

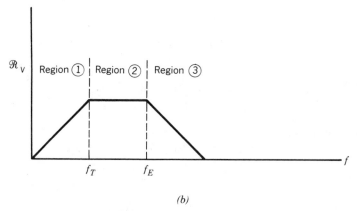

Figure 8-4 Responsivity frequency response: (*a*) electrical and thermal responsivities; (*b*) overall responsivity.

Now the voltage responsivity can be expressed:

$$\mathcal{R}_V = \frac{\omega p A_d R_d}{\sqrt{1 + \omega^2 \tau_E^2}} \frac{\varepsilon R_T}{\sqrt{1 + \omega^2 \tau_T^2}} \tag{8.13}$$

where the responsivity is a combination of thermal characteristics and electrical characteristics as identified in Eq. (8.13). A plot of voltage responsivity versus frequency can be divided into the electrical and thermal responses as shown in Fig. 8-4*a*. If one plots the thermal part of Eq. (8.13), the solid curve is formed with a 3-dB point at $f = 1/\tau_T$ (the

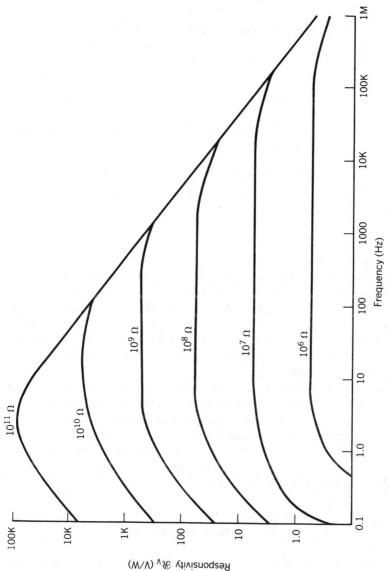

Figure 8-5 Typical responsivity $\mathscr{R}_V$ versus frequency for pyroelectric detectors with various external feedback resistors.

Table 8-2

Frequency Regions for the Pyroelectric Detector

Region	Responsivity
$f < \dfrac{1}{\tau_T}$	Proportional to frequency
$\dfrac{1}{\tau_T} < f < \dfrac{1}{\tau_E}$	Compensating rolloff from thermal and electrical causes a constant response
$f > \dfrac{1}{\tau_E}$	Inversely proportional to frequency

curves are drawn here for pedagogical purposes and are not to be taken literally). It rolls off as $1/\omega$ above this thermal time constant point. The electrical part of Eq. (8.13) is drawn dashed in Fig. 8-4a, it is increasing as ω until the frequency corresponding to the electrical time constant is reached, and then it is flat (independent of f) above this frequency. Therefore, the responsivity can be found by multiplying the electrical and thermal parts point-by-point. This produces three regions of interest in Figure 8-4b and Table 8-2.

The detector resistance (R_d) appears in Eq. (8.13). Actually, R_d should be replaced by the parallel equivalent of R_d and the load resistor (R_L). In fact the load resistance is much smaller than the detector resistance, so that the parallel sum equals the load resistance. A plot of responsivity $(\mathcal{R}_V)$ versus frequency can be developed for various load resistors as shown in Fig. 8-5. As the load resistor increases the voltage responsivity increases but the frequency response decreases.

8-3 NOISE SOURCES

There are four noise sources associated with a pyroelectric detector. They are temperature noise (radiation noise), Johnson noise of the detector and load resistor, and preamplifier noise.

8-3-1 *Temperature Noise*

The noise associated with temperature fluctuations, which was discussed in Chapter 2, is rarely the limiting noise for a pyroelectric detector, because pyroelectric detectors are seldom background radiation noise limited. If a detector is connected to a heat sink via thermal conductance K at a temperature T, it will attain thermal equilibrium to have zero mean power flow. However, the detector will exhibit a temperature fluctuation with equivalent thermal noise power

$$\Delta \phi_T = (4kT^2K)^{1/2} \tag{8.14}$$

This can be related to voltage fluctuations by the responsivity of the detector

$$\Delta V_T = \mathcal{R}_V \, \Delta \phi_T \tag{8.15}$$

8-3-2 *Johnson Noise*

This noise is usually the dominant noise source for pyroelectric detectors. In order to evaluate the Johnson noise of the detector an equivalent circuit must be evaluated to find the expression for resistance. Figure 8-6 shows

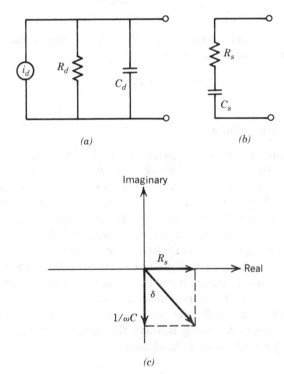

Figure 8-6 Equivalent circuits for a pyroelectric detector and phasor diagram: (*a*) parallel, (*b*) series circuit, (*c*) phasor diagram.

the equivalent circuit for a pyroelectric detector as a parallel and series circuit.

In the series circuit the pyroelectric detector can be thought of as a lossy capacitor. A phasor diagram as shown in Fig. 8-6c can be drawn for the resistance and capacitance of the series equivalent. The phasor angle δ is found from

$$\tan \delta = \frac{R_s}{1/\omega C_s} = \omega R_s C_s \tag{8.16}$$

This is called the "loss tangent" of the detector and can be used as a figure of merit. Ideally one would like a perfect capacitor with no series resistance ($R_s = 0$) and, therefore, a loss tangent of zero. In general, the lower the loss tangent, the better the detector will perform. The Johnson noise voltage associated with the series resistance can now be expressed as a function of frequency using this loss tangent expression

$$V_J^d = \sqrt{\frac{4kT \tan \delta \, \Delta f}{\omega C}} \tag{8.17}$$

where $\tan \delta / \omega C$ = series resistance, R_s. The Johnson noise voltage associated with the load resistor (R_L) is simply

$$V_J^L = \sqrt{4kTR_L \Delta f} \tag{8.18}$$

8-3-3   Preamplifier Noise

The choice of preamplifiers for a pyroelectric detector is very critical. It is desirable to be detector noise limited (loss tangent noise limited) and, therefore, the preamplifier must have lower input noise than the detector noise. There are both current and voltage noise sources at the input of the preamplifier. Figure 8-7 shows all the important noise sources for interfacing a pyroelectric detector to a preamplifier. Typically the current noise source can be eliminated if good junction field-effect transistors (JFETs) are used. Typical values for voltage noise for good JFETs are from 1 to 5 nV Hz$^{-1/2}$. The noise spectrum can be plotted as a function of frequency as shown in Fig. 8-8. Each region of the noise spectrum is due to a different noise source (Astheimer and Weiner). Note that there is no $1/f$ noise since no current is flowing through the detector. For region ① the load resistor's Johnson noise is the dominant noise, or

$$V_n^L = \sqrt{\frac{4kTR_L \Delta f}{1 + \omega^2 R_L^2 C^2}} \tag{8.19}$$

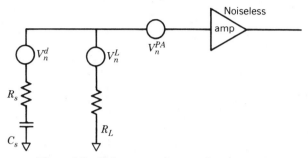

Figure 8-7 Noise sources for pyroelectric preamp.

which rolls off at

$$f_L = \frac{1}{2\pi R_L C} \tag{8.20}$$

where C is the total capacitance across the load resistor. The noise spectrum is down by 3 dB at f_L, but for simplicity, Fig. 8-8 contains only straight line segments. The Johnson noise rolls off above f_L at a $1/\omega$ rate of 6 dB/octave to frequency f_e. At this frequency the load resistor noise is equal to the loss tangent noise of the detector. So in region ② the noise follows the detector voltage noise expressions, which vary as the square root of frequency or 3 dB/octave:

$$V_n^d = \sqrt{\frac{4kT \tan \delta \, \Delta f}{\omega C}} \tag{8.21}$$

Finally, at frequency f_a, the preamplifier voltage noise and loss tangent noise of the detector are equal, so that the noise for higher frequencies is flat.

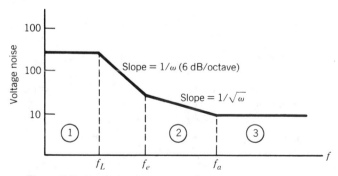

Figure 8-8 Typical noise spectrum for pyroelectric detectors.

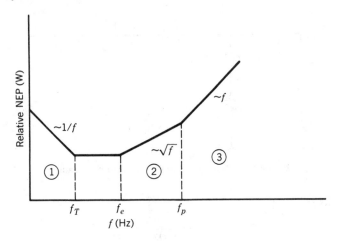

Figure 8-9 NEP versus frequency of pyroelectric detectors.

8-4 NOISE EQUIVALENT POWER

The noise equivalent power (NEP) as a function of frequency is simply the noise divided by the responsivity. For maximum responsivity, the shape of the responsivity versus frequency will be considered to be a triangle shape with the peak at f_e (recall Fig. 8-5). Therefore, by dividing this responsivity, which varies linearly with frequency to f_e, region ① can be drawn in Fig. 8-9 of the NEP response. For regions ② and ③ the responsivity is rolling off as $1/\omega$, so after dividing into the noise expression the NEP varies as $\sqrt{\omega}$ to frequency f_a, and linearly with ω above f_a. Typically, the flat region is from 10 to 40 Hz for the lowest possible NEP value. NEP values of 5×10^{-11} W have been achieved for 1 mm² detectors.

8-5 PREAMPLIFIERS

The preamplifiers used for pyroelectric detectors are of either the voltage or current-mode type. Figure 8-10 shows the circuit of both types. It should be clear from the load resistor/responsivity discussion and Fig. 8-5 that high load resistor values provide higher responsivity. This is true for either the voltage-mode or current-mode preamplifier. Since the resistance is typically high, JFETs are used to transform this high resistance to a low-impedance source. JFETs have large input resistance and low gate-current leakage. Load resistors on the order of 10^{11} Ω are used for maximum responsivity. One of the major problems associated with this

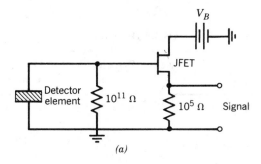

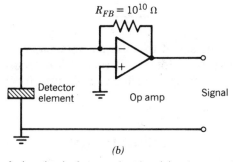

Figure 8-10 Interfacing circuits for pyroelectrics: (*a*) voltage mode, (*b*) current mode.

detector/preamplifier combination is microphonics. Therefore, the detector, load resistor, and JFET should be placed in very close proximity to each other.

8-6 CONCLUSIONS

Table 8-3 lists the currently used pyroelectric detectors with some parameters of interest.

The conclusions one makes are that the desirable characteristics of a pyroelectric detector are high Curie temperature, large pyroelectric coefficient, small dielectric constant, large resistance, low loss tangents, and small heat capacity.

The pyroelectric detector is very useful for infrared detection where cooling requirements are unacceptable or where sensitivity can be a trade-off for complexity. The pyroelectric vidicon is attractive in infrared

Table 8-3

Useful Detectors

Material	Curie (°C)	Pyroelectric Coefficient	Dielectric Constant	D^*
Tyriglycerine sulfite (TGS)	48	3×10^{-8}	25	5×10^8
Strontium barium niobate ($SrBaNbO_3$)	200	10^{-7}	8200	5×10^7
Lithium tantalate	600			2×10^8
Polyvinylidene fluoride (PVF_2)				10^8

imaging because of the signal-to-noise advantage over a single detector with scanning optics (see Chapter 5). Pyroelectric detector arrays with 256 elements in a single line are available, with two-dimensional imaging arrays in development.

BIBLIOGRAPHY

Astheimer, R. W. and S. Weiner, "Application Notes for Pyroelectric Detectors," Technical Bulletin 2-611, Barnes Engineering Company, Stamford, Conn.

Cooper, J., "A Fast-Response Pyroelectric Thermal Detector," *J. Sci. Instrum.* **39**(9), 467–572 (1962).

Cooper, J., "Minimum Detectable Power of a Pyroelectric Thermal Receiver," *Rev. Sci. Instrum.* **33**(1), 92–95 (1962).

Merz, W. J., "Double Hysteresis Loop of $BaTiO_3$ at the Curie Point," *Phys. Rev.* **91**(3), 513–517 (1953).

Putley, E. H., "The Pyroelectric Detector," *Semiconductors and Semimetals*, Vol. 5, R. K. Willardson and A. C. Beer (eds.), Academic Press, New York, 1970.

Roundy, C. B. and R. L. Byer, "Subnanosecond Pyroelectric Detectors," *Appl. Phys. Lett.* **21**(10), 512–515 (1972).

Roundy, C. B., R. L. Byer, D. W. Phillion, and D. J. Kuizenga, "A 170 psec Pyroelectric Detector," *Optic. Comm.* **10**(4), 374–377 (1974).

PROBLEMS

8-1.   Discuss the detection mechanism for a pyroelectric detector.

8-2.   What is loss tangent?

8-3. Draw a typical NEP versus electrical frequency plot. Explain each region.

8-4. Discuss the advantage of cooling a pyroelectric detector.

8-5. Discuss the effects of the atmosphere on thermal detector compared to a photodetector.

8-6. Why doesn't a pyroelectric detector work at zero frequency (DC)?

8-7. How is the optimum operating temperature determined for a pyroelectric?

8-8. A thermal detector ($\varepsilon = 1$ for all λ) is mounted on two posts to a heat sink. The steady-state heat flow to the heat sink via the posts is $1.0 \ \mu\text{W K}^{-1}$. Assuming equal thermal conduction in each post and one post breaks, then:

   a. The responsivity will increase by 2.
   b. The responsivity will decrease by 2.
   c. The responsivity will not be changed.
   d. The time constant will decrease by 2.
   e. The time constant will increase by 2.
   f. None of the above.

8-9. What can be done if the electrical time constant (τ_E) is larger than the thermal time constant (τ_T)?

8-10. Discuss a current mode of operation versus a voltage mode for a pyroelectric detector.

CHAPTER 9

CHARGE TRANSFER DEVICES

9-1 INTRODUCTION

The detectors discussed in the previous chapters have been primarily single-element detectors. However, it is desirable to extend these single elements into linear and two-dimensional arrays. Unfortunately, implementation of such arrays by the conventional means of adding individual detectors together becomes prohibitive for arrays of more than a few hundred detector elements. For example, focal-plane arrays using this type of construction would require a wire and probably a preamplifier for each detector. The result would be a maze of wires and processing electronics.

In 1970, Boyle and Smith presented the concept of charge-coupled devices (CCD), and later Amelio, Tompsett, and Smith (1971) developed the first operating device. With the use of a CCD, the detector signals could be retrieved without a maze of bonds. In addition, by using a CCD approach, the filling efficiency (photosensitive area to total focal plane area) became much larger.

The charge transfer device in the simplest form is a closely spaced array of metal–insulator–semiconductor (MIS) capacitors. The MIS was first suggested by Moll (1959) and Pfann and Garrett (1959). The most important is the metal-oxide-semiconductor (MOS) capacitor made from silicon and silicon dioxide as the insulator. Silicon dioxide is an insulator and is simply oxidized silicon. A detailed discussion of Si-SiO$_2$ MOS capacitors is given by Nicollean and Brews (1982) and Kim (1979).

A charge transfer device passes information between spatial locations by discrete packets of electronic charge. The charge position in the MOS

Figure 9-1 Photograph of a two-dimensional focal-plane array with 32×32 detectors (courtesy of Ford Aerospace).

array of capacitors is electrostatically controlled by voltage levels. Under proper application of these voltage levels and their relative phases, the MIS capacitor can be used to store and transfer the charge packet across the semiconductor substrate in a controlled manner.

Figure 9-1 shows a two-dimensional focal plane having a 32×32 array of dectors.

At the present time CCD arrays are available for the visible to the mid-infrared wavelength (0.4–20 μm). The arrays are either monolithic or a hybrid fabrication in which the detector material is sandwiched to a CCD array of cells.

This chapter discusses the basic MOS capacitors, focal-plane architecture and transfer techniques, noise analysis, and some commercially available arrays.

9-2 CONCEPTS OF CHARGE STORAGE

The basic building block of the charge transfer device is the metal-insulator-semiconductor (MIS) capacitor. A cross section of such a capacitor is shown in Fig. 9-2. A voltage V_g can be applied to the metal gate to charge or discharge the capacitor. The energy-band diagram for an ideal MIS capacitor is shown in Fig. 9-26 for a P-type semiconductor.

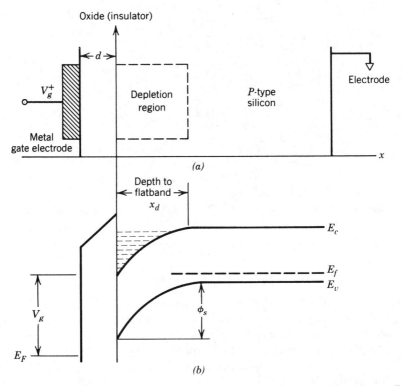

Figure 9-2 Metal-insulator semiconductor for a P-type substrate. Positive bias applied produces a depletion region: (a) structure of an MIS device, (b) corresponding energy-band diagram.

As shown in the figure the Fermi levels are shifted by the applied voltage V_g. The charge in the semiconductor under the gate is equal and opposite to that on the gate. In this ideal model the insulator has infinite resistance so no charge transport takes place through it.

The most important MIS capacitor to date is made from silicon because its natural oxide is an insulator. Although the use of germanium oxide is possible, the material lacks ruggedness and practicality. For applications in CCDs almost all MIS capacitors are metal-oxide-silicon (MOS) capacitors made from silicon, silicon dioxide, and metal (aluminum). For the remainder of this chapter we will discuss MOS capacitors exclusively.

Applying a positive voltage to the gate causes the mobile positive holes in the P-type semiconductor to migrate toward the ground electrode (like charges repel). This region, which has been depleted of positive charge, is called the depletion region, and its boundaries are shown in Fig. 9-2a. For the depletion case, a small positive voltage causes the bands to bend downward, and the majority of carriers are depleted in this region. The depleted region is modeled by what is called the "depletion approximation," which assumes that the edge of the depletion region is abrupt and that no mobile majority charge carriers exist in this depletion region.

To model this depletion region mathematically, it will be assumed that it is uniform but stops when the energy band becomes flat. This is known as the depletion approximation. It is illustrated in Fig. 9-2b as the depth to flatband. In this depletion region, electrons (minority carriers) can be accumulated and held against the insulator (oxide) surface.

The potential at the semiconductor/insulator interface is called the surface potential ϕ_s and is shown in Fig. 9-2b.

If a photon with energy greater than the energy gap is absorbed in this depletion region, it produces an electron–hole pair. The hole is mobile and moves out of the depletion region; however, the electron is held in the depletion region against the oxide interface.

The amount of negative charge that can be collected under the gate is proportional to the applied voltage as can be seen in Fig. 9-2 by the depth to flatband, but it is also dependent on oxide thickness and gate electrode area.

The charge distribution as shown in Fig. 9-3 can be substituted into the Poisson equation to solve for the potential cross section of the MOS capacitor

$$\nabla^2 \phi = \frac{\partial^2 \phi}{\partial x^2} = \frac{-\rho}{\kappa} = \frac{qN_a}{\kappa} \tag{9.1}$$

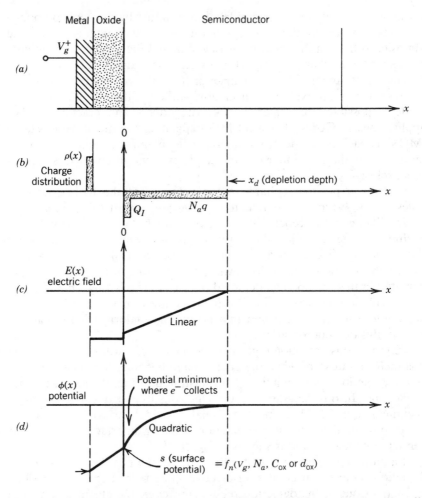

Figure 9-3 One-dimensional MOS model: (a) MOS cross section in one dimension; (b) charge distribution versus x; (c) electric field; (d) potential distribution versus x.

where ϕ = potential distribution,

x = distance,

ρ = charge density,

κ = dielectric constant,

q = electron charge,

N_a = acceptor doping concentration.

For the charge density distribution shown in Fig. 9-3 the electric field

E across the MOS structure can be solved by integration of $E = -\nabla \phi$:

$$\frac{\partial \phi}{\partial x} = \int_0^x \nabla^2 \phi \, dx = \int_0^x \frac{\rho}{\kappa} dx = \int_0^x \frac{qN_a}{\kappa} dx$$

As shown in Fig. 9-3, the field is constant across the insulator and is a linear function of x in the depletion region.

Upon one more integration (Fig. 9-3c to 9-3d), one can obtain the potential distribution as a function of x:

$$\phi = \int_0^x E(x) \, dx = \int_0^x \frac{qN_a x}{\kappa} dx$$

$$= \frac{qN_a x^2}{2\kappa} \quad \text{(in depletion region)}.$$

The potential varies linearly across the insulator and quadratically (x^2) in the depletion region. The boundary value at the gate, V_g, sets the limits. From Fig. 9.3d, one sees that the potential ($\phi_s = qN_a x_d^2/2\kappa$) has a minimum at the insulator/semiconductor interface, and thus the electrons accumulate at this minimum. This is called a surface-channel CCD capacitor. This minimum potential at the interface is the surface potential ϕ_s.

If equations for the potential distribution across the device and the charges are used, an expression can be found that relates the surface potential to the doping of the semiconductor, the gate voltage, and the oxide capacitance ($C_{ox} = \kappa/d$):

$$V_g = V_{ox} + \phi_s$$

where V_{ox} = voltage drop across the oxide.

The voltage drop across the oxide can be expressed in terms of the charge on a capacitor $V_{ox} = Q_g/C_{ox}$ and the charge on the gate, Q_g, by

$$Q_g = Q_I + Q_D$$

$$= qn_s + qN_a x_d$$

where Q_I = inversion charge density on oxide/semiconductor interface,
   Q_D = charge density in depletion region,
   n_s = number of electrons collected at interface (signal electrons),
   x_d = depletion depth.

Rewriting the equation for the potentials

$$V_g = \frac{qn_s + qN_a x_d}{C_{0x}} + \phi_s$$

or substituting for x_d as a function of ϕ_s gives

$$V_g = \frac{qn_s + \sqrt{2qN_a\kappa\phi_s}}{C_{ox}} + \phi_s$$

Solving for the surface potential ϕ_s, using the quadratic equation, yields

$$\phi_s = V_g + \frac{qN_a\kappa}{C_{ox}^2} - \frac{qn_s}{C_{ox}} - \sqrt{2\left(V_g - \frac{qn_s}{C_{ox}}\right)\frac{qN_a\kappa}{C_{ox}^2} + \left(\frac{qN_a\kappa}{C_{ox}^2}\right)^2} \qquad (9.2)$$

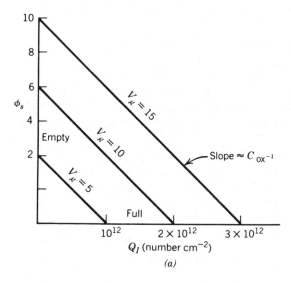

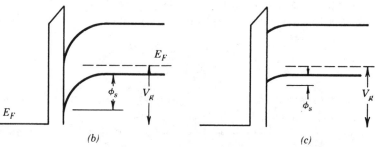

Figure 9-4 (*a*) Surface potential versus inversion charges for various gate voltages. Energy-band-diagram configuration for the (*b*) empty and (*c*) full well cases.

From Eq. (9.2), one sees that the surface potential varies with the doping concentration N_a and the oxide thickness, which affects C_{ox}. It is a maximum when no signal charge is collected ($n_s = 0$).

Figure 9-4a shows the variation of surface potential and inversion charge density Q_I for different gate voltages. The surface potential goes nearly to zero for a full well of charge.

The next section shows how an array of stored charges can be moved or transferred from its origin to adjoining locations for eventual transfer off the array.

9-3   CHARGE TRANSFER CONCEPTS

If an array of MOS capacitors is biased by the application of a proper sequence of clock voltage pulses, the charge packet can be transferred in a predictable manner through this array of capacitors.

Surface-charge transfer devices operate by transferring packets of minority charge from one MOS potential well (spatial region) to another. The value of the charge is proportional to the signal. These potential wells are formed by depleted MOS capacitors. The potential energy of these wells is determined by the voltage applied to the gate electrode and by the oxide thickness.

Since the charge moves to the spatial position of the adjacent minimum potential well, the charge will transfer along as the voltage levels on the gates are varied serially. Figure 9-5 shows two MOS capacitors with the depletion region under capacitor 2 larger than that under

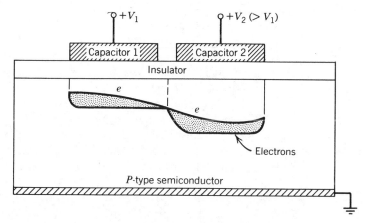

Figure 9-5 Two MOS capacitors biased for charge transfer to take place from 1 to 2.

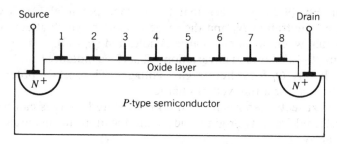

Figure 9-6 MOSFET with eight gates.

capacitor 1. The electrons have moved to the minimum potential as shown under gate 2. This is for an N-channel device that has a P-type semiconductor substrate.

Now consider a metal-oxide-semiconductor field-effect transistor (MOSFET) that is built with 8 gates instead of one as shown in Fig. 9-6. A charge is introduced at the source and is transferred to the drain by passing under gates 1 through 8. This can be done by serially biasing in time gates 1 through 8. The charge flows to the potential minimum as a positive voltage is sequentially placed on the gates and eventually comes out on the drain diode shown in Fig. 9-6. It is obvious that this approach can be extended to as many gates as desired. This is an example of a linear CCD array (one dimensional) with each voltage level independently controlled. If every fourth gate were connected together, the same process would occur but only four different voltage levels (phases) would be required.

A good mechanical analogy of the charge transfer mechanism is a machine consisting of a series of reciprocating pistons (Amelio, 1975) on a crankshaft as shown in Fig. 9-7 with metal shot (BB's) located at the lowest piston position. As the shaft controlling the piston position rotates in a counterclockwise direction, the BB's representing electrons are moved spatially from piston 3 to 4 then to 1 in half a revolution. This model also illustrates the loss mechanisms. Since the pistons have flat surfaces, not all the BB's are transferred from the previous gate.

A charge transfer device causes information to be passed between spatial locations in discrete packets of charge. In this chapter the term charge transfer device will be directed to devices where the charge positions in an MOS array of capacitors are electrostatically controlled.

If an array of MOS capacitors is aligned as shown in Fig. 9-8a, it forms a four-phase, N-channel device. The timing diagram (Fig. 9-8c) shows the voltage levels versus time for each phase, with specific times

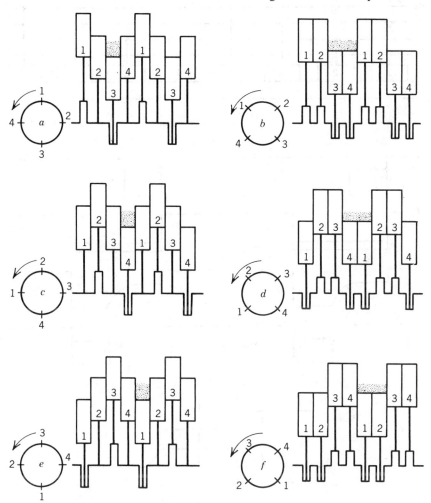

Figure 9-7 Mechanical machine analog to CCD operation.

t_1, t_2, t_3, and t_4 marked on the waveforms. By noting the well potentials below each gate at time t_1, one sees that the minority carriers (electrons) are collected under gate ϕ_1 since it is the minimum potential. At time t_2, when ϕ_2 has voltage applied, the charge packet begins its transfer to a location under gate ϕ_2. At time t_3, the voltage on gate ϕ_1 is collapsing and the charge is spilling to a new location under gate ϕ_2. At time t_4, the transfer is complete and the charge packet is now under gate ϕ_2. It has moved spatially one gate distance.

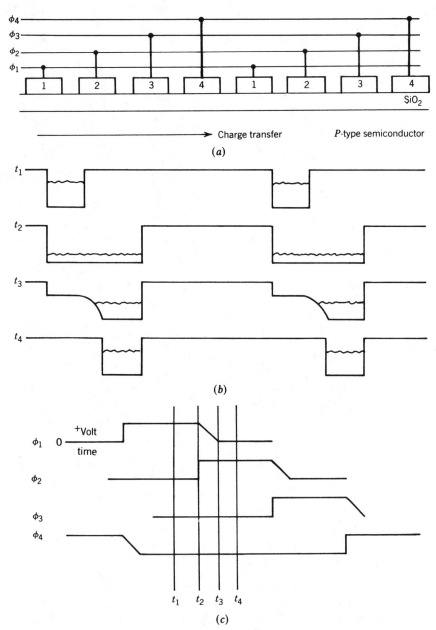

Figure 9-8 Four-phase system showing charge transfer. (*a*) This is a cross section of one column (rows go into paper). This column is isolated from adjacent columns by channel stop (P⁺). (*b*) Potential well profiles shown at different times as indicated on timing diagram. (*c*) Timing diagram of voltage waveforms applied to gates ϕ_1, ϕ_2, ϕ_3, and ϕ_4.

9-3-1 *Charge Transfer Mechanisms*

Charge is transferred from one spatial location (gate) to another by three mechanisms: (1) self-induced drift (repulsion of like charge); (2) thermal diffusion; and (3) fringe field drift. These mechanisms operate with a free charge transfer model, that is, free electrons in the conduction band (Carnes, Kosonocky, and Ramberg, 1972).

The charge transfer process is governed as in other detectors by the continuity equation

$$\frac{\partial n}{\partial t} = \frac{1}{q}\frac{\partial J}{\partial x} \tag{9.3}$$

where n = charge concentration,
$\quad J$ = current density,
$\quad t$ = time,
$\quad q$ = electron charge,
$\quad x$ = distance.

The current density is made up of self-induced-drift and thermal-diffusion contributions.

As a rule, self-induced drift accounts for about 95% of the charge transfer and thermal diffusion and fringe field drift account for the rest. The self-induced drift is caused by the mutual repulsion of like charges and is electrostatic in nature (Carnes, Kosonocky and Ramberg, 1971).

The expression for self-induced-drift charge density as a function of time is

$$N_s(t) = N_s(0)\left(1 + \frac{t}{\tau_{SI}}\right)^{-1} \tag{9.4}$$

where $N_s(0)$ = charge density at initialization transfer,
$\quad \tau_{SI}$ = self-induced time constant.

The self-induced time constant is

$$\tau_{SI} = \frac{2L^2 C_{ox}}{\pi \mu q N_s(0)} \tag{9.5}$$

where L = interelectrode spacing,
$\quad \mu$ = carrier mobility.

For an N-channel device with interelectrode spacing of 25 μm, C_{ox} = 1 pF, and charge density [$N_s(0)$] of 10^{12} states cm^{-2}, the self-induced time constant is 0.14 μsec.

The thermal diffusion of the charge follows the expression

$$N_s(t) = N_s(0) \frac{8}{\pi^2} e^{-t/\tau_{th}} \qquad (9.6)$$

where the thermal time constant, τ_{th}, is

$$\tau_{th} = \frac{4L^2}{\pi^2 D_n} \qquad (9.7)$$

where D_n = diffusion constant.

For $D_n = 10 \, \text{cm}^2 \, \text{sec}^{-1}$ and interelectrode spacing of 25 μm, $\tau_{th} = 0.25 \, \mu\text{sec}$. The high-frequency limit would be 4 MHz for this constraint. Since this time constant is so much larger, it will dominate the high-frequency transfer limit. (The lower frequency limit of operation is

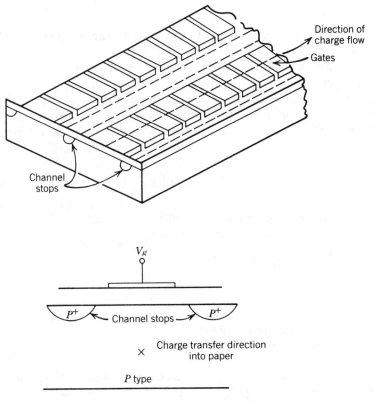

Figure 9-9   Channel stop configuration.

determined by dark current, which will be discussed in Section 9-6.) The effect of potentials on adjacent gates to which the well is emptying its charge is called fringing fields. These fields tend to smooth out the potential wells, which have been drawn as abrupt changes in the previous figures. These fringe fields tend to aid the charge transfer, especially when the thermal diffusion is dominant, so that the frequency response is higher than that predicted above by the thermal time constant alone. For further information on the fringe field effect the reader is referred to Krambeck (1971).

The transferred charge is forced to flow unidirectionally, that is, down a single column, by the use of channel stops. These channel stops are added to an array to avoid transverse charge migration. Figure 9-9 shows a cross section of a gate and the channel stops. The channel stops are P^+ regions that are formed either by a diffusion process or ion implantation. Since the surface potential (ϕ_s) is a function of doping [Eq. (9.2)], for high doping regions (P^+) the potential collapses and the surface potential is confined between the channel stops.

9-3-2 *Charge Transfer Efficiency*

The previous section indicates that the transfer mechanisms are not perfect; in other words, not all of the charge is transferred. Some signal charge is left behind after the transfer process. The term charge transfer efficiency (CTE) is the ratio of charge transferred to the initial charge present. A second term often used in the literature is transfer inefficiency, ε, or the ratio of the charge lost to the initial charge packet. Thus

$$\varepsilon = 1 - \text{CTE} \qquad (9.8)$$

Typical values of CTE are of the order of 0.99999 for a good device or an ε of 10^{-5}.

The effect of a nonunity CTE is to cause the signal packet to be spread to gates that are adjacent to the signal-carrying gate. The charge packets occur in what perhaps were empty wells that were following the signal charge and will reduce definition in the image. Figure 9-10 shows the effect on a packet of $100e^-$ for perfect transfer and for a CTE of 0.999 over 100 gates.

The relative amount of transfer for n transfers is $(1 - \varepsilon)^n$. For the preceding example, 90 electrons will be under the gate that is carrying the main charge. The other 10 electrons will be spread among the succeeding gates as shown in Fig. 9-10b. Typically, to measure the CTE for a device, a group of charge packets of equal amount must be transferred through the entire array of gates. The output packets of charge might appear as shown in Fig. 9-10c. To find the CTE, one then

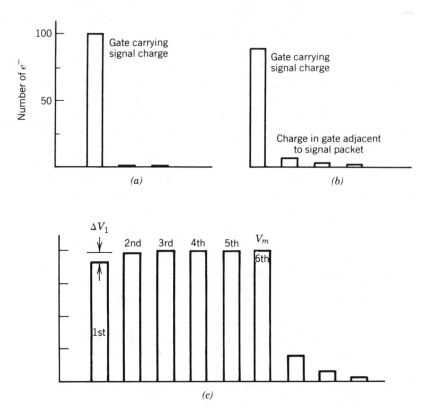

Figure 9-10 Effects of transfer inefficiency on signal charge: (*a*) Perfect transfer (CTE = 1.0) output after 100 transfers. (*b*) CTE = 0.999, output after 100 transfers. (*c*) Group of signal-carrying gates showing transfer loss.

uses the following equations for a good approximation:

$$(1 - \varepsilon)^n = \frac{V_m - \Delta V_1}{V_m} = (\text{CTE})^n \tag{9.9}$$

where V_m = maximum signal packet voltage,
  ΔV_1 = loss in lead pulse from V_m.

By rearranging the terms,

$$\text{CTE} = \exp\left[\frac{\ln(1 - \Delta V_1 / V_m)}{n}\right] \tag{9.10}$$

If n is large and the CTE is high, this can be approximated by the first two terms of a binomial expansion:

$$\text{CTE} \cong 1 - \frac{\Delta V_1}{n V_m} \qquad (9.11)$$

For our example in Fig. 9-10c, if $\Delta V_1 = 10$, $V_m = 100$, and $n = 100$, then CTE $\cong 0.999$ as we predicted.

Figure 9-10b shows that as the initial signal charge under gate 1 decreases, the charges under gates 2, 3, and 4 increase. However, all the charge is conserved; no real charge is lost, just rearranged.

The transfer mechanisms have already been discussed. Now, let us look at some of the loss mechanisms that cause imperfect charge transfer. The loss mechanisms that contribute to the charge transfer inefficiency are (1) incomplete free charge transfer in a clock period; (2) surface-state trapping; (3) bulk-state trapping; and (4) interelectrode gaps. A brief discussion of each of these transfer loss mechanisms will be presented to give a better conception of their origin.

For the case of incomplete free charge transfer in a clock period, the charge transfer efficiency can be written from Eq. (9.6) as

$$\text{CTE} = \frac{N_s(0) - N_s(t)}{N_s(0)} = 1 - \frac{8}{\pi^2} e^{-t/\tau_{th}} \qquad (9.12)$$

From Eq. (9.12) the CTE is a function of time. (Note we have assumed that the self-induced time constant and fringe field effects are negligible.)

Since the transfer is exponentially dependent on the time, the charge transfer cannot be perfect unless the clock period becomes very large, that is, as $t \to \infty$, CTE approaches 1.

A second loss mechanism is surface-state trapping. Trapping sites exist along the semiconductor–insulator interfaces that hold electrons momentarily. These electrons are momentarily captured and then reemitted with an emptying time constant τ_e of

$$\tau_e = (\sigma v N_c)^{-1} \exp(E_e/kT), \qquad (9.13)$$

where σ = captive cross section,
N_c = density of state,
v = thermal velocity,
T = temperature,
E_e = energy level of e^- from the conduction band,
k = Boltzmann's constant.

Therefore, the CTE is dependent on the clock frequency (time of

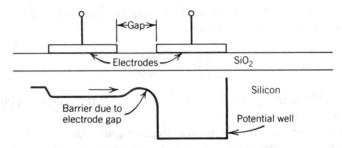

Figure 9-11   Potential barrier due to interelectrode gap.

transfer). If the trapped electrons emit quickly, they rejoin the main packet of charge. But they fall out of traps at a slower rate than they go into traps, thus they are left behind and the CTE goes down.

Bulk-state trapping is similar to surface-state trapping except the traps are within the substrate itself. These traps become important in buried-channel CCD arrays.

Interelectrode gaps also cause transfer inefficiency because the potential wells in the substrate do not monotonically decrease. As shown in Fig. 9-11, the gap produces a potential barrier over which the carriers must cross. This barrier impedes their flow and thus loss occurs.

Brodersen et al. (1975) classify the loss mechanisms discussed above into three basic types of charge transfer loss: fixed loss, proportional loss, and nonlinear loss. This approach lends itself to a mathematical model for transfer loss analysis.

Fixed loss is a loss of a fixed amount of charge during each transfer, and it is independent of the size of the signal charge. Proportional loss and nonlinear loss are dependent on the signal-charge level

$$N_L = \varepsilon N_s(0) \qquad (9.14)$$

where N_L = number of lost charges,

ε = transfer inefficiency,

N_s = number of signal carriers.

In the case of proportional loss, the transfer inefficiency ε is a constant and is independent of the signal-charge amplitude. In the case of nonlinear loss, ε is dependent on signal-charge amplitude.

The effect of a nonunity CTE on an image is to smear the image or reduce the modulation transfer function (MTF) of the array. If a sinusoidal signal is transferred through the array, the modulus is decreased and the

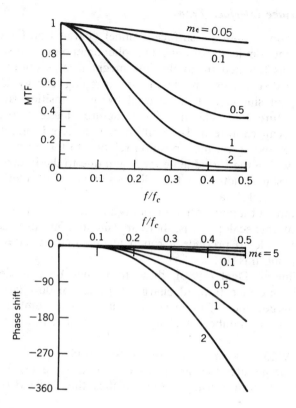

Figure 9-12 MTF and phase versus relative frequency response for various amounts of loss.

phase is shifted as given by (Sequin and Tompsett, 1975)

$$\text{MTF} = \exp\left[-m\varepsilon\left(1 - \cos\frac{2\pi f}{f_c}\right)\right] \tag{9.15}$$

$$\phi = -m\left[\frac{2\pi f}{f_c} - \sin\frac{2\pi f}{f_c}\right] \tag{9.16}$$

where m = number of gates the charge transferred,
f_c = cutoff spatial frequency response of the array.

The MTF and phase shift are plotted for varying amounts of charge loss, $m\varepsilon$, in Fig. 9-12. Therefore, it is desirable to have the largest MTF, and, therefore, to make the total loss ($m\varepsilon = \Delta V_1 / V_m$) as small as possible.

9-3-3 *Surface Interface Traps*

In this section we discuss surface-channel charge transfer devices and buried-channel devices. An important phenomenon of the surface-inter-face trap was touched on in the loss mechanisms section; however, it warrants further discussion. There are trapping energy levels in the energy gap of silicon that are caused by the transition from a silicon crystal structure to a silicon oxide structure. These trap sites in the energy gap can capture and reemit charge carriers in greater or lesser amounts, depending on their position relative to the Fermi level. These traps exist within the forbidden energy gap due to the interruption of the periodic crystal structure at the silicon–silicon oxide interface (Carnes and Kosonocky, 1972a).

An example of a conceptual model is an interface that is not perfectly flat on an atomic scale; that is, there are silicon atoms protruding up into the silicon oxide to produce a rough terrain that impedes the flow of electrons along this surface.

According to Deal (1980), these traps may be classified into four groups: (1) interface trapped charge; (2) fixed oxide charge; (3) oxide trapped charge; and (4) mobile ionic charge. The number of traps are expressed as the number of charges per unit area (number of charges cm^{-2}).

Figure 9-13 is a cross section of an MOS capacitor showing the various trapping sites that are possible. If one expresses the probability distribution of the trap using Fermi statistics, then for a *P*-type semicon-

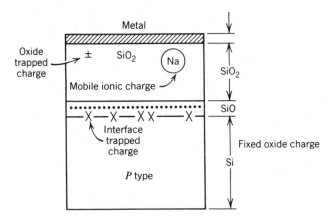

Figure 9-13   Trap associated with thermally oxidized silicon (from Deal, © 1980 IEEE).

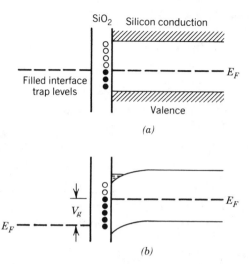

Figure 9-14 Trapping sites occupied as a function of gate bias: (a) Trapping states occupied with zero gate voltage. (b) Trapping states occupied with voltage V_g applied.

ductor

$$P(E_t) = \left[1 + \frac{1}{4}\exp\left(\frac{E_e - E_F}{kT}\right)\right]^{-1} \tag{9.17}$$

When a voltage is applied to the gate or electrode of an MOS capacitor, the valence and conduction bands bend as shown in Fig. 9-14. In Fig. 9-14b the energy levels of the interface traps cross over the Fermi level, providing for a higher probability of being occupied. Thus, more of these traps are filled by the signal charge.

During the charge transfer, the voltage applied to the gate causes the filled trap levels to be above the Fermi level, as shown in Fig. 9-14, and they lose electrons by reemission to the conduction band. When the trap level is below the Fermi level, it is occupied by electrons from the conduction band. As a charge packet enters under an electrode, the trap fills up. If the time constant is short compared to the transfer time, these levels will reemit into the main charge packet and not be left behind.

The rate of capture of electrons by these traps is given by (Singh and Lamp, 1976)

$$\frac{d\eta_{ss}(t)}{dt} = \underbrace{\frac{\sigma v n}{d}[N_{ss} - \eta_{ss}(t)]}_{\text{Capture rate}} - \underbrace{\eta_{ss}(t)\sigma v N_c e^{-E/kT}}_{\substack{\text{Emission time} \\ \text{constant}}} \tag{9.18}$$

where $\eta_{ss}' = $ density of filled surface state,
   $N_{ss} = $ density of empty surface state,
   $d_i = $ thickness of inversion layer,
   $\sigma = $ capture cross section,
   $v = $ thermal velocity.

This equation gives the rate of charge loss in a dynamic situation. The charge transfer inefficiency is (Sze, 1981)

$$\varepsilon \approx \frac{qkN_{ss}}{CV}\ln(P+1) \tag{9.19}$$

where $q = $ charge on an electron,
   $V = $ change in surface potential caused by signal charge,
   $C = $ capacitance of gate,
   $P = $ number of phases.

Surface-Channel Transfer Devices   To substantially reduce the effects of charge loss due to these traps in surface-channel charge transfer devices, the level of carriers being transferred in all wells should be constant to keep the interface traps filled at all times. If each potential well is empty ahead of the signal charge being transferred, then the signal charge will lose some charge to the gate surface traps for each transfer to an empty well, thus reducing the signal charge. If all wells have a small charge in them, these traps will be filled before the signal enters the gate, thereby reducing the fixed losses. However, the problem with introducing a constant amount of charge is that it reduces the dynamic range of the well.

This important technique of passing a constant charge through the potential wells to reduce the loss of interface traps is called the "Fat Zero" technique. The zero level of the CCD wells is simply increased above the zero-charge level. By filling the well about 10–20% (Fat Zero) before adding the signal, one will gain or lose charge from the trapping sites during each transfer and that is statistically a better situation (Beynon and Lamb, 1980).

The Fat Zero fills the surface states under the gate. However, when the signal is present, the area expands, exposing more surface states. This is called the "edge effect," and causes some loss in transfer efficiency, but this is a second-order effect. Figure 9-15 shows that the signal charge placed above the Fat Zero charge causes greater surface exposure $(A_T - A_{FZ})$; therefore, more interface trapping sites are exposed.

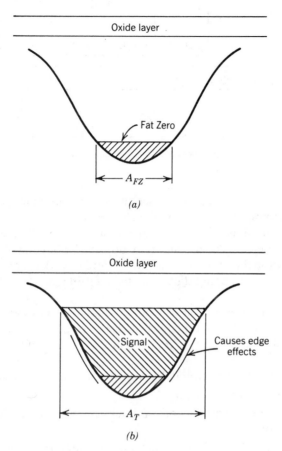

Figure 9-15   Fat Zero in potential well. (*a*) Potential well with Fat Zero. (*b*) Potential well with Fat Zero and signal charge, thus causing a greater number of surface traps to be exposed.

Buried-Channel Concept   A second way of reducing the effects of interface traps is to cause the signal charge to be moved away from the surface into the bulk material. This is accomplished by slightly doping the silicon substrate with a dopant of opposite polarity, that is, for an *N*-type substrate a *P*-type material can be ion implanted or diffused to about 1 μm. This type of device is called a buried-channel device (see Fig. 9-16). In this device the carriers are now majority carriers in the buried channel as opposed to minority carriers for the surface-channel case.

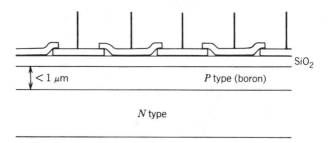

Figure 9-16   Buried channel CCD.

An analysis of a simple model of the buried-channel CCD MOS capacitor shows that the potential-well minimum (where the carriers tend to reside) will lie away from the silicon–silicon dioxide interface. The P region is modeled as being uniformly doped. We have avoided including the effects of metal and silicon having different work functions, the oxide charge, interface trapping states, and the built-in junction voltage (Vicars, 1982).

Consider a cross section of a buried-channel MOS capacitor that has four regions which are functions of distance x in our one-dimensional model, as shown in Fig. 9-17a ($x = 0$ at the P-type SiO$_2$ interface):

1. A metal gate at potential V_g with surface charge density Q_g.
2. An oxide layer of thickness, x_o.
3. A layer of P-type silicon of thickness x_P with uniform acceptor density of N_a.
4. A thick, N-type silicon substrate with uniform donor density N_d, connected to a ground at one end.

For this model of a buried channel, the P-Si region is held at a negative potential voltage of ϕ_c, as shown in Fig. 9-17a, sufficient to completely deplete it of holes. The P-Si region has a uniform space-charge density of $-eN_a$ as shown in Fig. 9-17b in the $x = 0$ to $x = x_P$ region. On the other side of this reversed-biased P-N junction located at $x = x_P$, a uniform positive space-charge density of eN_d extends a distance x_d into the N region. The distance x_d is determined by the requirement of electric charge neutrality that must exist across the structure. Hence, the charge placed on the gate Q_g determines the depletion depth x_d for the one-dimensional model:

$$Q_g = eN_a x_P - eN_d x_d \qquad (9.20)$$

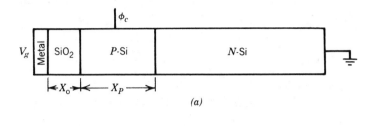

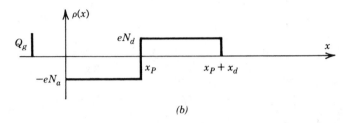

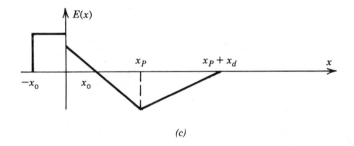

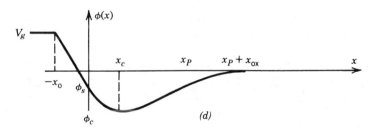

Figure 9-17 Buried-channel CCD storage cell: (a) structure, (b) charge density, (c) electric field, (d) potential.

Poisson's equation can be solved for the electric field for the various regions of the buried-channel MOS capacitor

$$\nabla^2 \phi = -\frac{\rho}{\kappa\epsilon} \qquad (9.21)$$

where ρ = charge density,
k = relative dielectric constant of that space, (silicon or oxide)
ϵ = dielectric constant of free space [8.854×10^{-12} F m^{-1}],
ϕ = potential.

Recall that the electric field is

$$E = -\nabla\phi \qquad (9.22)$$

By integrating the charge density for the three regions (see Fig. 9-17b) and using the boundary conditions, the electric field can be found as shown in Fig. 9-17c. In the oxide region, (k_o)

$$E(x) = \frac{Q_g}{\epsilon\kappa_0}, \qquad -x_0 < x < 0 \qquad (9.23)$$

In the P region,

$$E(x) = \frac{Q_g - eN_a x}{\kappa_S \epsilon}, \qquad 0 < x \le x_P \qquad (9.24)$$

where κ_S is the dielectric constant of the semiconductor. In the N depletion region,

$$E(x) = \frac{Q_g - eN_a x + eN_d(x - x_p)}{\kappa_S \epsilon}, \qquad x_P \le x \le x_P + x_d \qquad (9.25)$$

The electric field $E(x)$ is zero in the metal $(x = x_o)$ and in the N region beyond x_d. The profile of the electric field is shown in Fig. 9-17c.

The integration of x in the electrical field provides the potential levels in these various regions (integration of Fig. 9-17c is Fig. 9-17d)

$$\phi(x) = V_g - \frac{Q_g(x + x_o)}{\epsilon\kappa_0} \quad \text{for } -x_o \le x < 0 \qquad (9.26)$$

$$\phi(x) = V_g - \frac{Q_g x_o}{\epsilon\kappa_0} - \frac{1}{\kappa_S \epsilon}\left(Q_g x - \frac{eN_a}{2} x^2\right) \quad \text{for } 0 < x < x_P \qquad (9.27)$$

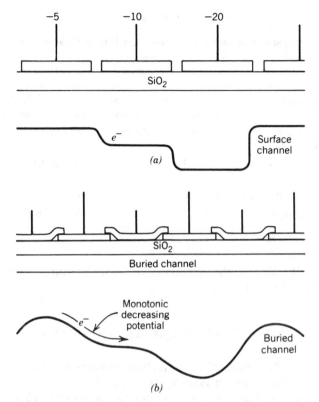

Figure 9-18 Potential profiles of surface-channel CCDs and buried-channel CCDs.

$$\phi(x) = V_g - \frac{Q_g x_o}{\epsilon \kappa_o} - \frac{1}{\kappa_S \epsilon}\left[Q_g x - eN_a x_P \left(x - \frac{x_P}{2}\right) + \frac{eN_P}{2}(x - x_P)^2 \right]$$

$$\text{for } x_P \leq x \leq x_P + x_d \qquad (9.28)$$

The buried channel, when coupled to other gates as shown in Fig. 9-18, produces a better potential gradient for charge to flow than does the surface-channel CCD. As shown in Fig. 9-18 the surface-channel CCD has flat potential areas where charge can be stationary; however, the buried-channel CCD has a continuous gradient for charge to flow to a potential minimum (Walden et al., 1972).

Some of the advantages of a buried-channel device are higher charge transfer efficiency, higher clocking frequencies, and lower surface traps. The disadvantages include lower signal-handling capability, more fabricating problems, and, currently, their inability to operate below 77 K.

Table 9-1

Comparison of Surface-Channel and Buried-Channel CCDs

Item	Surface Channel	Buried Channel
Charge storage capacity	4×10^{12} cm^{-2}, limited by oxide breakdown	10^{12} cm^{-2}, limited by silicon breakdown
Charge transfer efficiency	Requires Fat Zero	No Fat Zero
Clocking frequency limit	15 MHz	25–300 MHz
Fabrication complexity	Relatively simple	Additional masking step
Cryogenic operation	Yes	Not presently below 77 K

Table 9-1 highlights the comparisons between surface-channel and buried channel CCDs.

9-4 FOCAL PLANE ARCHITECTURE

So far we have been talking about the basic building blocks of an array and how signal-charge transfer takes place within the array. In this section we will extend those ideas into how a two-dimensional array (area array) can be made using MOS capacitors, and the various means of reading out the information. Ideally, one would want an array that is sensitive over 100% of its area. This sensitivity, that is, focal-plane filling efficiency, is defined as the ratio of the photoactive area of the detector array to the total area of the array. Since the capacitors are isolated islands as shown in Fig. 9-19

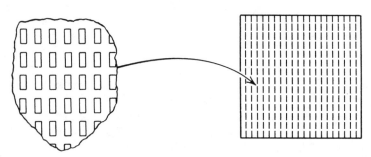

Figure 9-19 Typical focal-plane-array filling efficiency.

and require interconnecting leads, 100% filling efficiency is desired but is not possible.

Figure 9-20 shows one way to interconnect the gate electrodes on the capacitors. The timing diagram of this four-phase system for ϕ_1, ϕ_2, ϕ_3, and ϕ_4 is the same as that shown in Fig. 9-8c for transferring the charge down the columns. In this configuration a 1000-detector array can be operated with only 16 leads from the focal plane. As each row is shifted

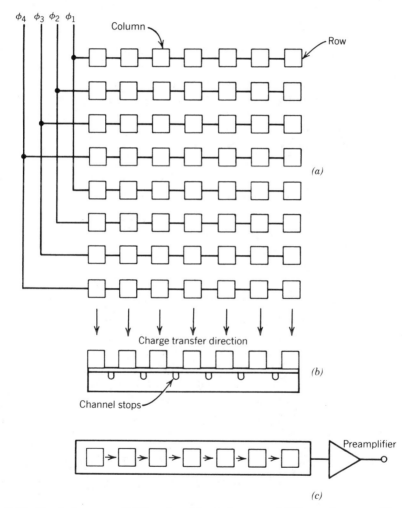

Figure 9-20 Focal-plane array: (a) Top view of focal-plane-array MOS capacitor gates. (b) Cross section of gates. (c) Output shift register.

down, the charge is placed on an output shift register that transfers the charge in the horizontal direction at a much higher speed to the output amplifier as shown in Fig. 9-20c. An on-chip amplifier converts the charge to a voltage. The conversion efficiency of this amplifier is expressed in volts per electron; typically about $1\ \mu V/e^-$ is a rough order of magnitude value.

A focal-plane assembly made up of several CCD arrays is called a mosaic. Several CCD arrays are butted together to form a larger focal plane. However, because of the necessary access to the output pads, additional layout problems occur in fabricating arrays for mosaic focal planes.

The arrays discussed so far are classified as monolithic arrays because the photosensitive detector material is an integral part of the array; that is, both the photosensing process and the charge transfer process are performed on the same substrate. There are other architectural arrangements where the detector material and the CCD are separate chips that are bonded together to form an array. Basically, there are three approaches to focal plane architecture:

1. Take a material (silicon) that has known signal-processing characteristics and learn to make a detector with it.
2. Take a well-established detector material (HgCdTe) and a well established signal-processing material (silicon) and hybrid them.
3. Take a well-established detector material and learn to make signal-processing CCDs with it (InSb).

Hybrid arrays are most important in infrared systems where infrared sensing materials (e.g., HgCdTe) are not presently compatible with CCD techniques. In this case a detector array of HgCdTe may be bonded to a silicon CCD multiplexer to process the signal information. More will be said about this in Section 9-4-4.

There are basically three schemes for reading data from a focal-plane array. They are called (1) line address; (2) frame/field transfer; and (3) interline transfer. Each scheme has a certain integration time to take an optical image and convert it into an electron distribution in the CCD wells.

The simplest form of array is called the line address (see Fig. 9-21). One pixel from each column is shifted to the output multiplexer (MUX), and this entire row is clocked off the array before a second pixel is shifted to the output multiplexer (MUX). Consequently, the output MUX must be clocking much faster than the column. If there are four columns, then

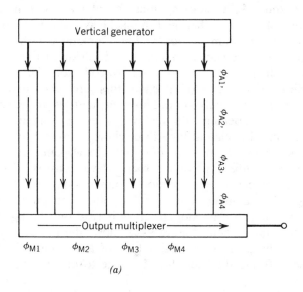

(a)

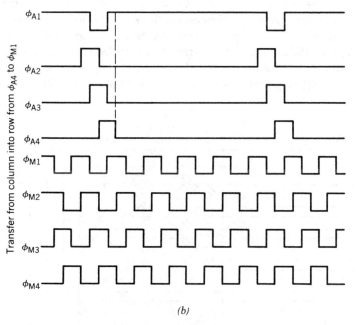

(b)

Figure 9-21 Line address layout: (*a*) array transfer sequence; (*b*) timing diagram.

the output MUX clocking must be at least four times as fast as the column clocking so that the last pixel leaves the MUX before the transfer of another column of pixels. The clocking for the array phases and output MUX is shown in Fig. 9-21b. Prior to this readout phase, the gates, which are biased-on, act as collectors for the photogenerated carriers. Since the MOS capacitors are photodetectors, the filling efficiency is relatively good. However, there is image smearing since the readout is happening with the image present. This smearing is minimized by causing the readout time to be short compared to a frame time.

The second means of reading out a focal-plane array is the frame/field transfer. In this approach a focal plane consists of two sections with separate clocks for each section. One section is used to detect the optical image during integration and the other section is used to store the previous image. The storage array is read out during the integration of the next image field. Figure 9-22 shows the layout of this type of focal plane. The problem with this type of architecture is that the image is present during the short transfer time to the storage area, thus causing image smear.

The interline transfer scheme is shown in Fig. 9-23. Vertical readout registers are interlaced with the photosensitive sites, which are arranged in columns. After collecting photoelectrons during some integration time (called a field), the charge is transferred to a shielded (aluminum-covered) shift register by a pulse on the transfer gates. The signal charge is now in the column shift registers as shown in Fig. 9-23 and is clocked out while

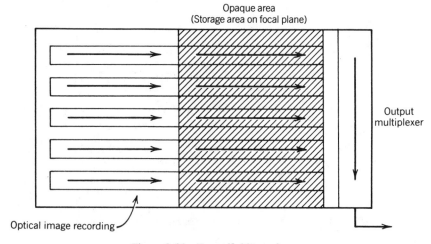

Figure 9-22   Frame/field transfer.

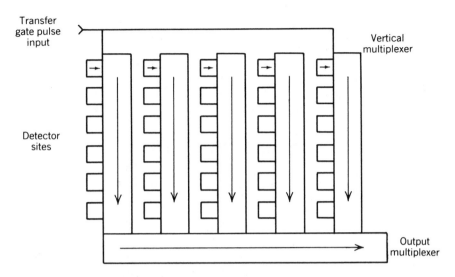

Figure 9-23 Interline transfer.

the next field is being integrated. For this type of device the same photosites are used for each field, thus, there is no interlacing of fields as with conventional TV unless special construction and clocking are done. However, there is no image smearing since the gates used to transfer the signal off the array are shielded. In addition, this type of architecture increases the integration time since the detector sites are not used to transfer the information off the array. There is, however, a loss in filling efficiency in this interline transfer type of array, owing to the requirement for auxiliary transfer shift registers.

One of the advantages of a CCD focal plane should be pointed out at this time. The amount of data that are produced can be reduced by simply changing the time sequence of the array. For example, as shown in Fig. 9-24 several pixels (nine in this example) can be summed together on the focal plane before the entire signal is transferred out. This binning process occurs when a single gate in the array register has a charge dumped into it and added algebraically. Then a similar three pixels in the output multiplexer are summed under one gate. The signal addition is shown in Fig. 9-24 as voltage versus time. This superpixel then represents nine pixels. If a signal of interest appears in a certain area, the option of zooming in electronically is then possible.

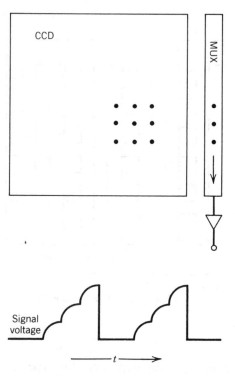

Figure 9-24   Data-compression technique.

9-4-1   *Discussion of Clocking Techniques*

We have restricted our discussion to four-phase (4ϕ) charge-coupled devices. However, CCDs can be extended to any number of phases, such as, for example, the three-phase system shown in Fig. 9-25 in which the charge is isolated on both sides by gates. The disadvantage of the three-phase CCD is that it requires three masking steps in the fabrication as opposed to two for the four-phase CCD. Figure 9-26 shows the layout of a four-phase CCD with two overlapping gates; all the electrodes associated with any one phase have the same mechanical orientation and configuration. Such is not the case for a three-phase array. Consequently, this mechanical difference causes potential-well nonuniformities to exist under the gates of a given phase across the array.

In the case of a two-phase CCD array, our present model would not work, since the charge would simply move back and forth without a specific direction as the gate voltages toggled. Nevertheless, two-phase arrays are important because there are two fields per frame, which are

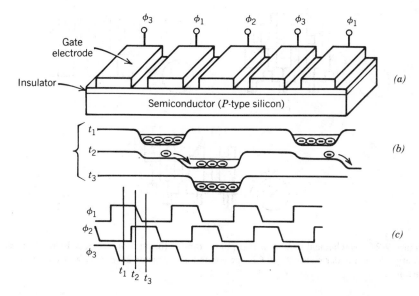

Figure 9-25 Three-phase CCD layout and operation (illustration courtesy Fairchild): (*a*) Mechanical layout of device. (*b*) Charge packet transfer sequence. (*c*) Timing diagram for 3ϕ transfer.

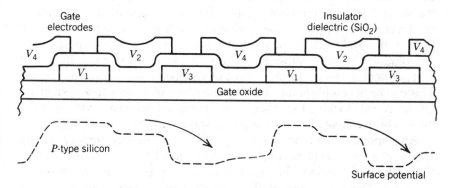

Figure 9-26 Four-phase clocking layout.

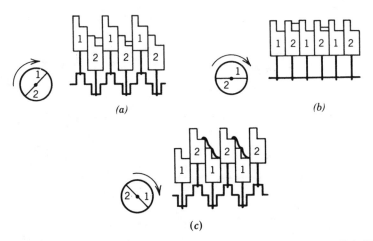

Figure 9-27 Mechanical analog of two-phase transfer (illustration courtesy Fairchild): (*a*) Charge located under piston 2. (*b*) Charge located under piston 2. (*c*) Charge transferred to piston 1.

compatible with commercial television and broadcasting systems. Therefore, it was desirable to create a two-phase CCD array. To make the charge unidirectional, a scheme was developed as shown in the mechanical analog in Fig. 9-27 (Amelio, 1975). Under each gate a potential well was established that was asymmetrical, thus causing the charge to be transferred in one direction. This asymmetric potential well can be produced either with a stepped oxide or a diffusion implant, as shown in Figs. 9-28*a* and 9-28*b*. The oxide thickness or the heavy doping causes the variation in potential wells to cause charge flow to be unidirectional.

The simplest system would consist of a single phase to drive the device. This can be accomplished by biasing one phase of a two-phase system at a DC level and clocking the other phase above and below it as shown in Fig. 9-29. This is classically called a phase and a half (Younse, 1982).

An important approach to a single phase is the development of the virtual-phase CCD (Hynecek, 1981). A virtual phase is constructed inside the bulk silicon below the oxide level and is biased to the substrate potential. The advantage of the virtual phase is that it eliminates the problem of shorts between phases during the fabrication of multiphase systems.

The phase could be a single polysilicon sheet across the array. However, to increase the quantum efficiency and improve photon absorption, the gates are laid out in a conventional manner. Since there is

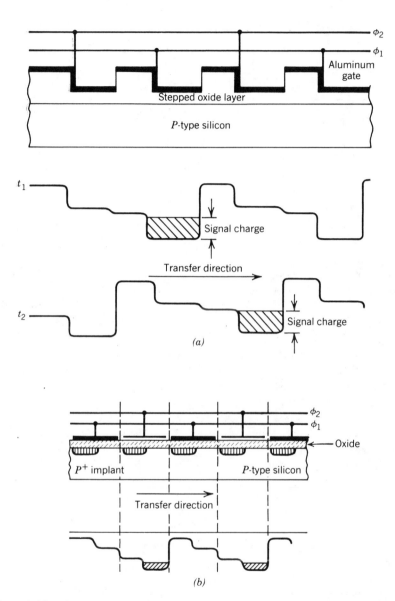

Figure 9-28 Two-phase CCD architecture: (a) Two-phase CCD using stepped oxide; (b) Two-phase CCD using diffusion implant.

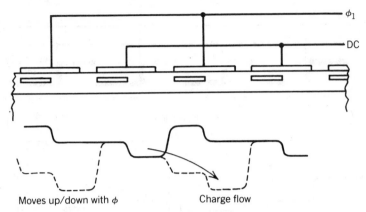

Moves up/down with ϕ Charge flow

Figure 9-29 One-phase CCD operation. One set of electrodes has a dc level of half the clock driving amplitude of ϕ_1.

only one phase, shorts between gates, which are a major fabrication problem, are not an operational problem.

The trade-offs in using the various types of clocking are primarily dependent on the application. However, other considerations are also important. Two-phase clocking has the least number of transfers. Three-phase clocking dissipates the least amount of power and has the highest filling efficiency (most pixels for a given area of chip). Four-phase clocking has the largest signal-handling capability, which is important for infrared systems. This added signal-handling capability is due to the way the gates are clocked. Figure 9-30 exemplifies two clocking wave forms, normal and double clocking (Barbe, 1975). In the case of double clocking, the charge is transferred under two gates simultaneously, thus increasing the charge-handling capability by a factor of 2. Table 9-2 shows the various clocking trade-offs one can make (Barbe, 1975).

Table 9-2

Clocking Trade-offs (Barbe, © IEEE 1975)[a]

	2ϕ	3ϕ	4ϕ
Pixel length	$2L$	$3L$	$4L$
Driver power dissipation	$4\,CV_g^2 f_g$	$3CV_g^2 f_g$	$4CV_g^2 f_g$
Charge-handling capacity	$1\times$	$1\times$	$2\times$
Transfers per pixel	2	3	4

[a]L = minimum feature dimension along channel; f_g = clock frequency; C = capacitance.

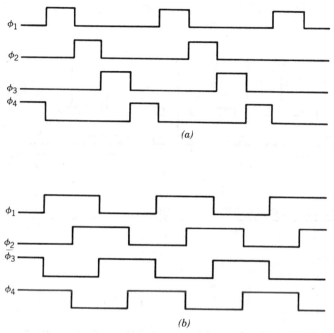

Figure 9-30   Two clock waveforms for four-phase clocking (after Barbe, © 1975 IEEE): (a) normal and (b) double.

Since commercial TV is so important, the sequence of events for operating a standard TV system will be detailed for a two-phase array with interlacing fields:

1. Applying a positive voltage (V_A) to all ϕ_1 electrodes.
2. While V_A is applied (1/60 sec) the $h\nu$ is producing charge carriers under the electrodes—in potential wells (integration time).
3. Transfer ϕ_1 pixels out along the channels—in about 1000 μsec.
4. Apply voltage to ϕ_2 electrodes and integrate for 1/60 sec.
5. Transfer ϕ_2 pixel elements out in about 1000 μsec.
6. Now we have two interlaced fields which form a standard TV picture directly in 1/30 sec.

9-4-2   *Front or Back Illumination*

The direction from which radiation is incident on an array has considerable effect on its spectral responsivity. In front illumination optical energy

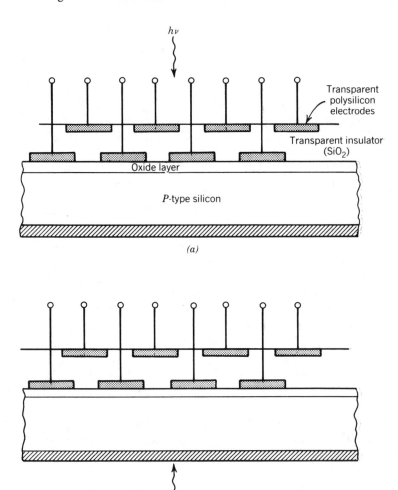

Figure 9-31 Illumination orientation: (*a*) front illumination and (*b*) back illumination.

is incident on the same side of an array as the CCD gates; in back illumination the photons are incident on the opposite side of the array (see Fig. 9-31) The problem with front illumination is the loss of detector response caused by interference and absorption effects at the electrodes (see Fig. 9-32). In the case of polysilicon electrodes, these effects degrade the blue response of a silicon-type detector.

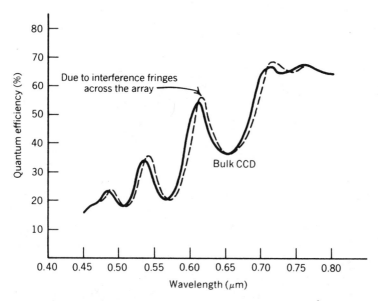

Figure 9-32 Quantum efficiency versus wavelength of RCA CCD (40 Å passband, 100 Å stops) (after Fowler et al., 1981).

Back illumination is a possible way to eliminate some of these problems. However, the transmission of silicon must then be taken into account. For infrared applications, back illumination can be advantageous since all the visible photons are filtered out. However, for a silicon-type array that responds from 0.3 to 1.1 μm, the substrate must be thinned down so that the distance to the depletion region is within one diffusion length of the carrier so that it can be detected. For back illumination the modulation transfer function (MTF) and blue sensitivity are better for thinner arrays. However, for ruggedness and red response, a thicker array is desired. The thickness of these back-illuminated arrays used for the intrinsic response (0.3–1.1 μm) varies from 10 to 30 μm (Shortes, 1974).

To further improve the sensitivity of back illumination, one could use an antireflection coating on these devices.

9-4-3 Monolithic Focal Plane Arrays

A monolithic focal plane is one in which the detector and the CCD are made in the same chip and not as a sandwich fabrication. Figure 9-33 shows an example of back-illuminated monolithic array. Table 9-3 lists several monolithic arrays of which silicon is the most common.

Table 9-3

Monolithic Focal-Plane Arrays

Material	Operating Temperature (K)	Spectral Response (μm)
Intrinsic		
InAsSb	100	2–8
HgCdTe	40–77	2–15
InSb	77	1–5.5
PtSi	77	1.1–5
Silicon	300	0.3–1.1
Extrinsic		
Si:Ga	18	2–16
Si:In	40	2–8

Monolithic arrays can be further categorized into monolithic extrinsics such as Si:In, Si:Bi, and Si:As and monolithic intrinsics such as HgCdTe, InSb, and InAsSb.

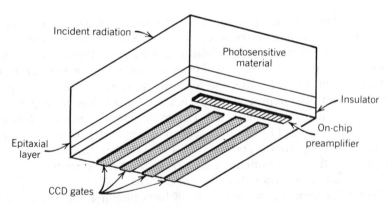

Figure 9-33   Monolithic array of a single material.

9-4-4 *Hybrid Focal Plane Array*

In the case of a hybrid focal plane, photosensing is done with one material (chip) and the CCD readout is done with another chip, normally silicon.

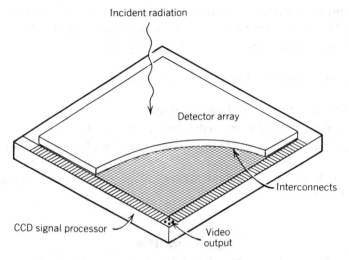

Figure 9-34 Hybrid-focal-plane array.

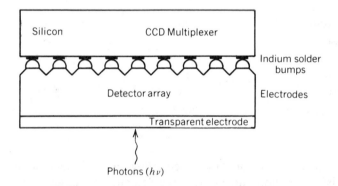

Figure 9-35 Hybrid-focal-plane bonding techniques.

An example of a hybrid array that sandwiches two chips together is shown in Fig. 9-34. The bonding technique used is a flip-chip solder bump bond as shown in Fig. 9-35. Indium is usually used to solder the detector array to the CCD multiplexer. This construction allows for higher filling efficiency in backside illumination.

The advantages of a hybrid architecture are:

1. Independent optimization of the detector material and CCD multiplexer.

2. Universal modular readout—a single CCD/output circuit for many different detector materials (i.e., different spectral responses).
3. Increased signal-processing area on the focal plane.

The disadvantages of the hybrid approach are:

1. Many mechanical and electrical interfaces.
2. Thermal expansion mismatch limits the size of the array.
3. Input circuit into CCD multiplexer must be efficient and noiseless.

Table 9-4 lists some hybrid arrays, all of which use a silicon CCD multiplexer.

Table 9-4

Hybrid Focal Planes

Material	Operating Temperature (K)	Spectral Response (μm)
InAsSb	100	2–8
InSb	77	1–5.5
HgCdTe	40–77	2–14
Si:In	42	2–8
Si:Ga	14	2–16
Pyroelectric	300	1–30

For the hybrid focal planes, a coupling circuit or technique is required to transfer the photocurrent from the detector to charge in the CCD well. There are three requirements for this coupling: (1) high efficiency; (2) low noise; and (3) simple architecture for space ("real-estate") limitations of unit cell. These input coupling circuits can be divided into direct and indirect injection.

1. Direct injection—the photocharge goes directly into well.
2. Indirect injection—voltage-to-charge conversion:
   a. fill and spill input;
   b. diode cutoff (gate modulation) input;
   c. capacitance metering input;
   d. floating diffusion input.

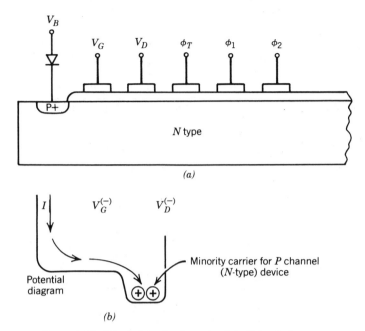

Figure 9-36   Direct injection input for hybrid focal plane.

Direct Injection   For this type of input of a hybrid focal plane, the photovoltaic detector is directly bonded to a diffusion on the CCD devices as shown in Fig. 9-36a. The gates V_G and V_D are held at a negative voltage to produce a depletion condition as shown in Fig. 9-36b, which causes the charge to flow directly under V_D. The transfer gate, ϕ_T, is held at zero potential during the integration time. It is then driven negative by a pulse as is ϕ, so the charge under the V_D gate is transferred into the CCD clocking gates. This approach is good for high-impedance detectors only, where the charge is photogenerated. For low-impedance devices the current is not photocurrent, but is due to intrinsic carrier concentration, and the potential wells fill up very rapidly making this approach undesirable.

Indirect Injection: Fill and Spill   Indirect injection has two important features: (1) it is linear, and (2) it has low noise.

This type of input as the name indicates is a two-step process. Charge is filled and then is spilled. An analogy is that if one wants to get a cup of water (exactly), one can fill the cup over the brim and the excess water will spill out. In this way one will always get a cup of water.

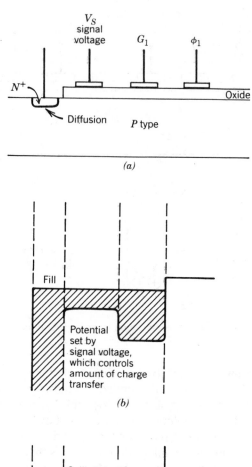

(a)

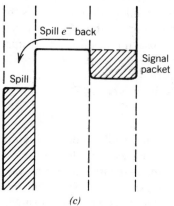

(b)

(c)

Figure 9-37 Fill and spill input.

230

The layout of the gates is shown in Fig. 9-37a. A voltage level that is proportional to the signal is applied on the gate marked V_S. During the filling process the diffusion is forward biased thus causing the electrons to come to the surface as shown in Fig. 9-37b. Gate G_1 is held low so that the electrons can fill the entire potential well as shown during forward bias of the diffusion (Kosonocky and Carnes, 1975).

When the input diffusion is back biased as shown in Fig. 9-37c, the electrons all spill except for the electrons trapped by the barrier formed by V_S. There is a packet of electrons captured under G_1 that is proportional to the barrier $(V_S - G_1)$. this packet of charge is

$$q = C_{ox}A(G_1 - V_S) \tag{9.29}$$

Indirect Injection: Diode Cutoff Indirect injection can be accomplished by a diode cutoff scheme as shown in Fig. 9-38. The diffusion (N^+) is an infinite source of electrons contained in the conduction band; the potential applied controls the level of this infinite source as shown in Fig. 9-38b.

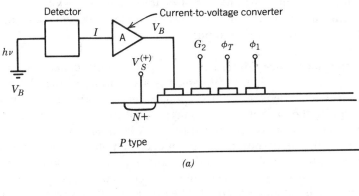

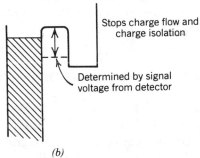

Figure 9-38 Diode cutoff input.

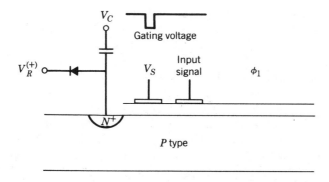

Figure 9-39 Capacitance metering.

A voltage level proportional to the detected signal is applied at V_G. V_G determines whether the diode (N^+/P type) input has current flowing (diode cut-on or cut-off). As shown in Fig. 9-38b, the depletion level under V_G determines the amount of charge transfer to storage gate G_2. After some integration time, the transfer gate is pulsed, ϕ_T, and the charge is transferred to the output CCD, ϕ_1.

Indirect: Capacitance Metering An input scheme called capacitance metering is shown in Fig. 9-39, which is a variation on the fill and spill approach. The diode is set positive, V_R^+, so the N^+ diffusion does not spill charge under the ϕ_1 gate. The voltage signal level is set on gate V_S as shown. A negative pulse is applied to the N^+ diffusion via the capacitor, which pulls the diode potential well above V_S potential well and simultaneously ϕ_1 has a voltage applied to it which produces a potential well to collect and hold electrons. This filling process continues until the input diode and the input gate are equal. The problem with this type of approach is that it is electrical time constant (RC) dependent and the charge packet is not repeatable.

9-5 OUTPUT READOUT TECHNIQUES

The output devices associated with a CCD focal plane can also be called on-chip preamplifiers. The purpose of the preamplifier is to sense the signal charge coming out of the CCD registers and to convert it to a usable signal. Consequently, the preamplifier has problems similar to those associated with the input circuits for a hybrid array: low noise, linearity,

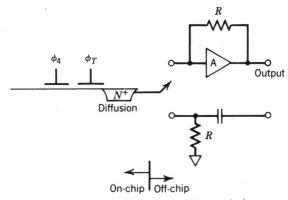

Figure 9-40   Current sensing (destructive).

compactness, and clock independence, that is, it is unaffected by clock feedthrough.

The output preamplifier does not depend on whether a monolithic or hybrid array is used. The charge can be sensed destructively or by nondestructive readout. The destructive-readout processes are either current sensing or floating diffusion amplification. The nondestructive-readout methods are: (1) charge sensing, (2) floating gate amplifier, and (3) distributed floating gate amplifier. A brief description of each of these will be given for completeness, however, the destructive readout of the floating diffusion amplifier is most commonly used. In current sensing (destructive) the charge is transferred off the CCD array onto a diffusion as shown in Fig. 9-40 via a transfer gate (ϕ_T). The diffusion is connected to a resistor or a transimpedance amplifier that converts charge to voltage.

This type of readout is linear, low in noise, and relatively decoupled from clocks. This preamplifier may have an additional DC gate before the transfer gate. The disadvantage of this approach is that the stray capacitance is high, and therefore it is limited to low-frequency operation.

9-5-1   *Floating Diffusion Amplifier (Destructive)*

The on-chip output preamplifier is shown in Fig. 9-41 along with the timing of the reset phase and phase ϕ_4. When ϕ_4 goes low, the electrons are transferred to the output gate node through the screen gate (V_{SCR}). The screen has a low DC voltage to allow transfer only when ϕ_4 goes off. The reset gate (RG) was set to the positive drain voltage (V_D) previously by the reset gate (ϕ_{reset}) pulse as shown in Fig. 9-41. The electrons

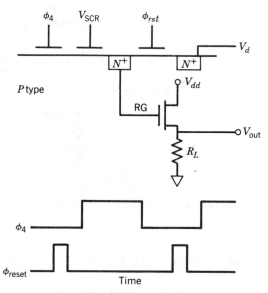

Figure 9-41 Floating diffusion amp (destructive readout). V_d, V_{dd}, and V_{SCR} are DC levels.

transferred to the floating diffusion cause the node to go negative toward ground and, simultaneously, the source follows. The output signal is the difference between the reset level and the negative-going pulse. To get the signal (ΔV_{FD}), V_{out} is typically differentially sampled after reset and after charge transfer, (correlated double sampling) and is digitized in a fast analog-to digital converter. Sometimes a gate is placed between ϕ_{reset} and the diffusion to reduce clock feedthrough.

In this case the preamplifier is on the chip itself. The signal ΔV_{FD} is the product of the signal charge (Q_{sig}) placed on the floating diffusion capacitance (C_{FD}) and the gain of the circuit:

$$\Delta V_{FD} = \frac{Q_{sig}}{C_{FD}} \times \text{Gain} \qquad (9.30)$$

If we rewrite the gain for the circuit shown in Fig. 9-41, we obtain

$$\Delta V_{FD} = \frac{Q_{sig}}{C_{FD}} \frac{g_m R_L}{1 + g_m R_L} \qquad (9.31)$$

where g_m = transconductance.

9-5-2 *Charge Sensing Amplifier (Nondestructive)*

An electrode is put into the oxide layer, which is connected to an output
transistor and is capacitively coupled to a fixed gate bias V_B (see Fig.
9-42). As the charge is transferred along the transfer gates, the charge
passes under this floating electrode, which senses the charge, $V = Q/C$.

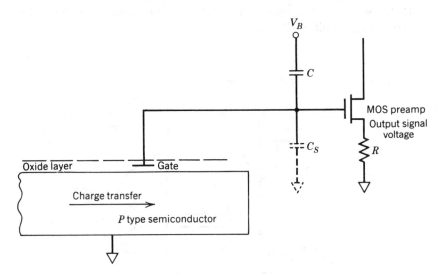

Figure 9-42 Charge-sensing output.

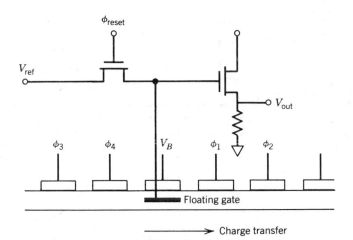

Figure 9-43 Conventional floating gate amplifier output circuit for a CCD (Barbe, 1975).

9-5-3 *Floating Gate Amplifier (Nondestructive)*

Figure 9-43 shows the configuration for the floating gate amplifier type of readout. A gate is in series with the phase transfer gates that senses the charge as it moves past. However, this gate is connected to a reference voltage through a reset gate and an output floating gate. The reset gate sets the floating gate potential to V_{ref}, and, as charge is passed under the gate, the voltage is reduced by $V = Q/C_{FD}$. The change in voltage is the signal voltage (Milton, 1980).

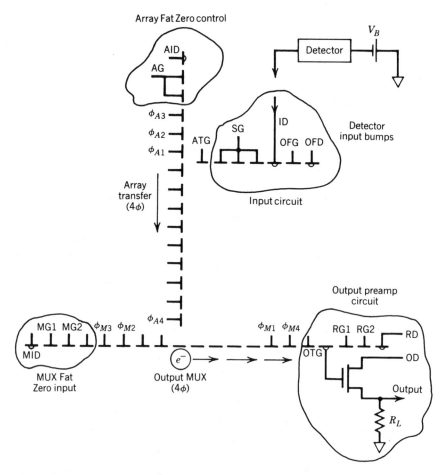

Figure 9-44a Gate layout indicating charge transfer with fill and spill input and floating diffusion output.

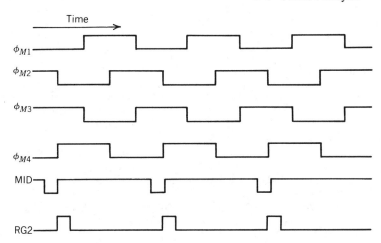

Figure 9-44b Timing diagram for output MUX with Fat Zero input. V_{sig}, V_{DC}, V_{ref}, V_{TET}, V_{SCR}, V_{dd}, and V_{d2} are all DC levels.

9-5-4 *Distributed Floating Gate Amplifier (Nondestructive)*

Since the readout from the floating gate scheme is nondestructive, the charge can be sensed at several sites as it moves along. This sampling of the same charge from several floating gates is called distributed floating gate amplification (DFGA) (Barbe, 1975), and it is used to improve the signal-to-noise ratio by the square root of the number of gates.

An example of a typical hybrid CCD circuit is appropriate to illustrate the charge input, transfer, and output techniques for a CCD array. This example has the most commonly used input/output technique, that of a fill and spill input and a floating diffusion amplifier output. Figure 9-44 shows a linear CCD cross section of gates and diffusions that are used to get charge into the CCD, transfer it, and measure the charge at the output. The timing diagram illustrates the voltage waveform for each gate to accomplish this transfer. The voltage levels all have equal amplitudes in the figure, in general however, they all are different values. DC voltage levels are also necessary as called out in Fig. 9-44.

9-6 NOISE ANALYSIS

There are a number of noise sources that limit the sensitivity of a system using a CCD focal plane. There are noises associated with the scene photon flux, noises related to the CCD itself, and noises related to the

Table 9-5

Types of Noise in CCD Focal-Plane Array

Photon noise	This manifests itself as shot noise due to the random arrival and emission rate of photons. Approximated by a Poisson process that results in a standard deviation equal to the square root of the mean number of photons.
Input noise	Random injection of charge from a diffusion into a potential well. Often called Fat Zero noise when a Fat Zero is put in electrically.
Transfer inefficiency noise	The noise associated with the random amount of charge lost by a signal upon transfer and the amount of charge introduced to a signal upon entering a well.
Trapping noise	A noise arising from random trapping emission from interface states or bulk states.
Dark-current noise	Also called thermal generation noise. This noise is associated with carriers that are thermally generated to bring the potential well into thermal equilibrium.
Clock feedthrough noise	This noise is due to capacitance coupling from the array gates to the output diode. It becomes more important at high frequencies.
Floating diffusion reset noise	This is a noise associated with the reset circuit on the output preamplifier. This noise is the thermal noise of the MOSFET channel resistance in parallel with the floating diffusion capacitance, often called kTC noise.
Amplifier noise	This noise is associated with a MOSFET of a given transconductance.
Detector uniformity noise	This noise is the variations across the video output for a uniform radiation flux input. It can be caused by responsivity variation or dark-current generation being spatially varying.
Read noise	Noise associated with reading the information from the focal-plane array—independent of time between reads.

Table 9-6

Expressions for Noise Sources Associated with Focal-Plane Arrays

Type	Expression rms Carrier Fluctuation	Terms
Photon	$[2G^2 \eta E_P A_d T_i]^{1/2}$	G = photoconductive gain η = quantum efficiency E_P = photon irradiance A_d = pixel area T_i = integration time
Input	$[kTC_{in}q^{-2}]^{1/2}$	k = Boltzmann's constant T = temperature C_{in} = input capacitance q = charge of electron
Transfer inefficiency noise	$[2\varepsilon N_s]^{1/2}$	ε = transfer inefficiency N_s = number of carriers in signal packet
Trapping noise	$[MkTA_d N_{ss} \ln 2]^{1/2}$	M = number of gate transfers N_{ss} = density of surface states
Dark current	$\left[\dfrac{J_d \tau_i A_d}{q}\right]^{1/2}$	J_d = dark current density τ_i = integration time
Floating diffusion	$[kTC_0 q^{-2}]^{1/2}$	C_0 = output capacitance
Preamplifier noise	$\left[\dfrac{C_0^2}{q^2} \dfrac{8kT\Delta f}{3g_m}\right]^2$	Δf = electrical bandwidth g_m = transconductance

output preamplifier. These noise sources are shown in Fig. 9-45. As is the case with any system, the best possible performance is obtained when the noise is caused by the randomness of the photons arriving at the detector. For the purpose of categorizing the noise sources, Fig. 9-45 depicts the focal-plane array as consisting of three sections. Table 9-5 lists the various noise sources and gives a brief description of each noise type (Carnes and Kosonocky, 1972b). The expressions for these various noise sources are given in Table 9-6. The noise is expressed in number of carriers, that is

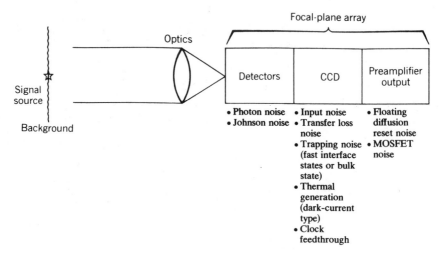

Figure 9-45   Noise sources in a system using a CCD.

100 electrons of noise. Some typical magnitudes of the various types of noise are given in Table 9-7.

Recall that the independent sources of noise add in quadrature, so to obtain the total noise one must root sum square all the noise sources.

Table 9-7

Typical Noise Level of Various Contributors at Room Temperature

Noise Source	Level (electrons)
Input noise	500
Transfer inefficiency	50
Trapping	
Surface channel	1000
Buried channel	100
Dark current	100
Floating diffusion reset	500
Amplifier	500

9-6-1 *Photon (Shot) Noise*

This noise is due to the random fluctuation in the number of photons incident on the detector. Since this process is stochastic with Poisson statistics determining the photon arrival times, the noise is equal to the square root of the average number of photons. This noise manifests itself as shot noise in the device. recalling the shot noise expression of

$$\overline{i_n^2} = [2eI\Delta f]^{1/2} \tag{9.32}$$

where I is the photogenerated DC current and is equal to $\eta G E_p A e$. The terms are defined in Table 9-6. A factor of 2 is added because generation and recombination of charge carriers in photoconductors are present. With photovoltaic detectors, omit the factor of 2 since recombination is not present and omit the photoconductive gain (G) since it is 1:

$$n_p = [2\eta E_p A_d \tau_i]^{1/2} \quad \text{(number of electrons)} \tag{9.33}$$

9-6-2 *Input Noise*

The noise associated with transferring charge onto a capacitor is often called kTC noise. This is the thermal noise of a resistor in parallel with a capacitor as shown in Fig. 9-46 (Beynon and Lamb, 1980).

The Johnson noise associated with the resistor R is

$$i_J^2 = \frac{4kT\Delta f}{R} \tag{9.34}$$

By simple network analysis one can show that the noise current through the capacitor is

$$i_c = \sqrt{\overline{i_J^2}} \frac{R}{R + 1/j\omega C} = \frac{j\omega RC}{1 + j\omega RC} \sqrt{\overline{i_J^2}} \tag{9.35}$$

where $1/j\omega C$ is the reactance of the capacitor.

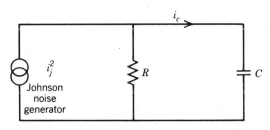

Figure 9-46 Equivalent circuit of input of charge to a potential well.

The total amount of charge noise present can be found by integrating the noise power spectrum of this noise current over all frequencies to give the mean square charge,

$$\overline{\Delta Q_n^2} = \int_0^\infty Q_N^2(\omega)\, d\omega \tag{9.36}$$

$$= \frac{1}{2\pi} \int_0^\infty \left[\frac{\tau^2}{1 + \omega^2 \tau^2} \right] \frac{4kT}{R}\, d\omega \tag{9.37}$$

where $\tau = RC$
$d\omega = 2\pi\Delta f$

Substituting a variable of integration for $x = \omega\tau$, so $dx = \tau\, d\omega$. Equation (9.37) can be rewritten as

$$\Delta Q_N^2 = \frac{4kT\tau}{2\pi R} \underbrace{\int_0^\infty \frac{dx}{1 + x^2}}_{\pi/2} = kTC. \tag{9.38}$$

This is the mean square charge across the capacitor producing noise. It is important to note that it is independent of the resistor value and only depends on temperature and capacitance.

There is a trade-off between noise and dynamic range (minimum capacity), since they both depend on C.

Another simpler way to obtain this result is to think of the energy in a capacitor and the minimum energy for a single degree of freedom to be in equilibrium:

$$\frac{Q^2}{2C} = \tfrac{1}{2}kT \tag{9.39}$$

$$Q^2 = kTC \tag{9.40}$$

The number of noise electrons is related to the charge noise by the charge on an electron ($q = 1.6 \times 10^{-19}$ C). The number of noise electrons associated with the input to a CCD can be expressed as

$$n_I = [kTCq^{-2}]^{1/2} \quad \text{(number of electrons)} \tag{9.41}$$

9-6-3   Transfer Inefficiency Noise

If N_S is the signal carrier in the signal packet, then on the average εN_S will be left behind at each transfer (Barbe, 1980; Thornber, 1974). There will be a fluctuation about this mean with a mean squared value of $2\varepsilon N_S$.

The factor of 2 is introduced because of the shot noise introduced once for entering and a second time while leaving the potential well:

$$n_\varepsilon = [2\varepsilon N_S]^{1/2} \quad \text{(number of electrons)} \tag{9.42}$$

9-6-4 Trapping Noise

Another noise associated with the transfer of charge from one site to another is caused by trapping or emission of fast interface states. These fast interface states are important noise contributors in surface-channel CCD (Sequin and Tompsett, 1975). If one assumes that these trapping states have reemission time constants on the order of the transfer time, an expression for this noise can be found.

For a surface-channel CCD (SCCD) the fractional rate of emission $e(E)$ of trapped carriers from a given interface state of energy E is given by the Shockley–Read–Hall equation as

$$e(E) = \sigma(E)\bar{v}_{th}N_c \exp(-E/kT) \tag{9.43}$$

where E = energy of that state relative to the nearest band edge
$\sigma(E)$ = capture cross section
$\bar{v}_{th}$ = mean thermal velocity of the charge carriers
N_c = the density of states in the band under consideration.

The probability P of a carrier trapped at energy level E being emitted in time t is given by

$$P = 1 - \exp[-te(E)] \tag{9.44}$$

Thus for a small energy band ΔE of interface state and a certain area A_d per electrode over which the interface states are filled, the distribution of full and empty states at time t will be binomial with a variance, $\overline{\Delta Q_{ss}^2}$, given by

$$\overline{\Delta Q_{ss}^2} = q^2 P(1 - P) A_d N_s(E) \Delta E \tag{9.45}$$

Using the previous equations and integrating over all the possible energy levels in the energy gap, see Fig. 9-14:

$$\Delta Q_{ss}^2 = \int_0^{Eq/2} q^2 N_s(E)(\{1 - \exp[-te(E)]\} \exp[-te(E)]) \, dE$$

$$= q^2 kT A_d N_{ss} \ln 2 \quad \text{for } t, N_s(E) \text{ constant} \tag{9.46}$$

This is for one transfer; therefore for M transfers, a multiplier of M is needed. In addition, this mean squared charge must be converted to electron number. The noise for a total fo M transfers due to fast interface states under complete charge transfer is

$$n_T = [MkTA_dN_{ss} \ln 2]^{1/2} \qquad (9.47)$$

Even in the complete charge transfer model there are other limitations on the charge transfer efficiency caused by trapping. The trap states exist in both surface-channel and buried-channel devices; however, the density of states (N_{ss}) in the surface channel is about two orders of magnitude higher than the buried channel devices.

The rate of filling these traps depends on the population in the conduction band, and the rate of decay of the trap state depends on the energy difference between the conduction band and the trap level. Therefore, there is a net charge pile up since the trapping states fill faster than they empty. This loss can be reduced as discussed earlier by introduction of a Fat Zero. This Fat Zero keeps the fast interface states continuously filled so that no traps are empty to capture charge from the signal packet.

9-6-5   *Floating Diffusion Reset Noise*

In this type of readout, the charge packet is transferred to a floating diffusion, which is connected to a gate of an output MOSFET. Basically the charge packet is dumped onto a capacitor to produce an output voltage. The noise is again the kTC type of noise discussed earlier (Milton, 1980).

A technique known as correlated double sampling can be used to suppress reset noise as well as reset level shifts. Low-frequency noises (such as 60 Hz) are smoothed by this high-frequency sampling technique. Correlated double sampling is described in Section 9.7 under Noise Measurements.

9-6-6   *Preamplifier Noise*

The output MOSFET has noise associated with it and has been shown to be (Gordon and Gordon, 1965):

$$n_A = \frac{C_0^2}{q^2} \frac{8kT\Delta f}{3g_m} \qquad (9.48)$$

9-6-7   *Detector Uniformity Noise*

This noise is associated with the output from pixel to pixel, or the variations across the entire array. It has a pattern such as that shown in

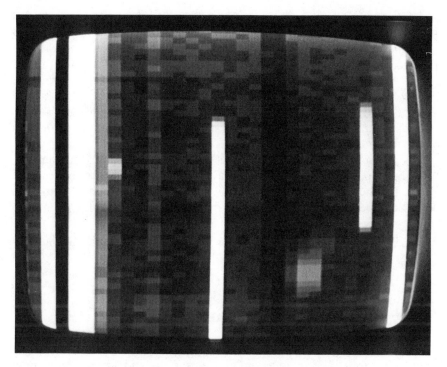

Figure 9-47 Photograph of display with fixed pattern noise.

Fig. 9-47. The checkerboard appearance is caused by a combination of two effects: (a) fixed pattern noise and (b) responsivity variations.

Fixed Pattern Noise Fixed pattern noise is a nonuniformity appearing across the array which does not vary with irradiance. Several effects contribute to fixed pattern noise. Thermally generated carriers are not uniform across the array. Local defects in the semiconductor crystal lattice cause a variation in dark-current generation across the array.

In the case of hybrid arrays, the input coupling for each detector may vary. This is more commonly called threshold nonuniformities. In any case, the individual detectors see a different input impedance to the CCD array. Thus, even if the detector array is perfectly uniform, the input coupling produces fixed pattern noise.

The last major contributor to pattern noise is the errors in the fabrication process. The errors introduced by inaccurate pixel areas caused by masking misalignments can lead to 10% nonuniformities.

The fixed pattern nonuniformity can be compensated using a storage and subtraction technique. The array response in the absence of optical radiation is stored (ususally in digital memory) and is subtracted electronically from all image frames produced by exposure to optical radiation. An apparently uniform response is generated using this procedure.

Responsivity Variation   Responsivity variation is a nonuniformity which has a variable effect as a function of irradiance. The responsivity of each pixel is determined by its local stoichiometry. Therefore doping variations across the device produce pixel to pixel responsivity variations.

The responsivity of each individual pixel can be measured to provide a basis for the correction of responsivity variations. The output from each pixel can be multiplied by the appropriate factor such that it appears to have the average responsivity of the array. This apparent uniformity can only be maintained at the expense of reducing the average dynamic range of the array. Most of the pixels probably will be well below saturation when the most responsive pixel saturates.

9-7 CCD ARRAY TESTS

There are two general optical–electronic arrangements for operating a detector array: (1) an imaging optical system and an image readout electronic system that form a camera system, and (2) a radiometric test chamber and a versatile set of test electronics that can be used for performance and diagnostic tests. This test equipment is much more complicated and more versatile than that required for normal camera operation (Dereniak et al., 1981).

A block diagram of a system that could be used for testing an array is shown in Fig. 9-48. The tests that can be performed on the detector arrays are:

1. Spectral responsivity.
2. Quantum efficiency.
3. Linearity and dynamic range.
4. Charge transfer efficiency.
5. Noise.
6. Noise equivalent photon irradiance.
7. Conversion efficiency.

These tests are performed after preliminary operations have verified

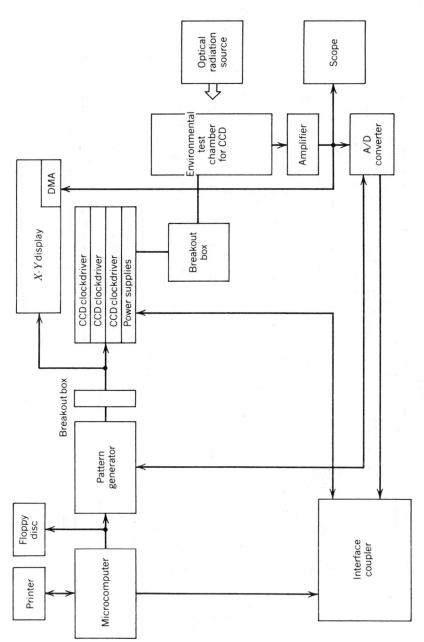

Figure 9-48 Charge transfer device testing facilities.

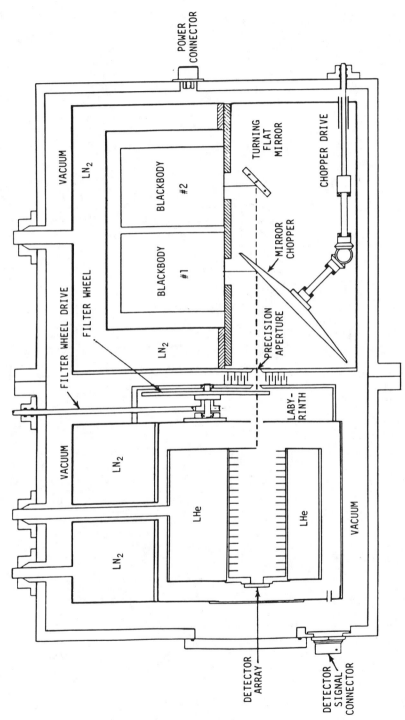

POWER CONNECTOR

VACUUM

LN₂

BLACKBODY #2

BLACKBODY #1

TURNING FLAT MIRROR

CHOPPER DRIVE

MIRROR CHOPPER

FILTER WHEEL

FILTER WHEEL DRIVE

LN₂

PRECISION APERTURE

VACUUM

LN₂

LABY-RINTH

LN₂

LHe

LHe

VACUUM

DETECTOR ARRAY

DETECTOR SIGNAL CONNECTOR

Figure 9-49 Environmental test chamber.

that the detector array is operating correctly. The equipment necessary to operate the array includes the pattern generator and charge transfer clockdrivers shown in Fig. 9-48. These devices produce the waveforms that are needed to correctly move the charges off the array in an orderly manner. Any focal-plane array can be tested as long as the correct timing pulse sequence is known and inserted into the pattern generator. In the setup shown in Fig. 9-48, a microcomputer is used to process the different timing pulse sequences for the different focal-plane arrays. Once these timing sequences are developed, they can be stored on a floppy disc for future recall by the microcomputer. In addition, the microcomputer is used for preliminary data reduction. The video signal from the array is amplified by a variable gain amplifier, and then it is processed by an analog-to-digital (A/D) converter. The A/D converter converts each pixel signal into a digital signal that is stored in the microcomputer. Special hardware that can store several frames of data and process the video signal by several different digital-image-enhancement techniques such as frame-to-frame subtraction or addition is commercially available. This special equipment is used primarily to remove fixed pattern noise.

There is a great deal of current interest in testing infrared CCD (IRCCD) arrays. For IRCCD testing, the most important part of the test system is an environmental test chamber. It houses the detector array, filter assembly, and radiation sources. A layout of an environmental test chamber is shown in Fig. 9-49. This chamber consists of two individual chambers bolted together. By separating the two chambers and using auxiliary cover plates (not shown) each chamber can be operated independently.

The right-hand chamber contains two blackbody sources and a reflective chopper. The temperatures of the blackbodies are determined by means of platinum resistance thermometers inserted in the cores. The thermometers also serve as feedback sensors for temperature controllers. The elevated temperatures of the two blackbodies are provided by heater windings on the cylindrical cores. The cores are insulated from their exterior casings by multilayer film insulation. Power and sensor connections are made by means of one or more hermetic connectors in the chamber wall. The chopper is mounted on a bearing-supported shaft and driven through a rotary vacuum feedthrough in the chamber wall. A liquid-nitrogen canister surrounds the two larger blackbodies and terminates in a cold plate upon which the blackbodies rest. A cold shield connected to the cold plate surrounds the chopper and turning flat. This cold shield also supports the precision aperture, which provides the primary stop of the optical system.

The exterior face of the cold shield holds a series of concentric rings centered on the aperture. These rings interleave with a similar set of rings on the left-hand chamber and provide a labyrinth radiation shield for the aperture, preventing 300 K radiation from entering the optical system.

The left-hand chamber contains a tandem set of cryogenic reservoirs. The upper reservoir, which is connected to the cold shield, always contains liquid nitrogen when operating. The lower reservoir may hold the cryogenic liquid of choice—nitrogen, helium, or neon. This lower reservoir contains a central tube that allows infrared radiation to pass through the lower reservoir from the right-hand chamber to a cold plate upon which the detector array is mounted. The interior of the tube contains a set of regularly spaced baffle rings that prevent stray radiation from reflecting into the detector array. A filter wheel in the optical path is externally positioned, so that various discrete and linearly variable spectral filters may be set in front of the system aperture. An auxiliary lens attachment (not shown) may be installed over the array so that the rediation can be focused on discrete pixels. This lens array consists of four small zinc selenide lenses of short focal length in a square pattern. The lenses are laterally positionable over limited distances so that selected pixels may be irradiated. Power and signals to and from the detector array are brought out through a hermetic connector near the array in the Dewar wall. An adjacent wall port permits access to the array and its interconnecting wiring. Interconnections to the detector array consist of fine gauge constantan wire to reduce heat leakage.

The seven tests or parameters to be measured on the arrays as mentioned earlier are best discussed in relationship to the capabilities of the environmental test chamber. Since the parameters must be evaluated for various operating voltages and temperatures, the large amount of data handling requires computer processing.

1. *Spectral Responsivity.* The chamber is outfitted with a circular variable filter that is cooled to liquid-nitrogen temperature. It is rotated by a stepper motor to scan the spectrum from 1.1 to 8 μm, using various circular variable filters.

2. *Quantum Efficiency.* The quantum efficiency η, which is defined as the ratio of free photogenerated carriers to incident photons at the detector surface, should be measured. The photon flux (photon sec^{-1} cm^{-2}) at the detector surface will be known to within ±5%. The errors introduced by the photolithography of the pixel size ($\Delta x + \Delta y$) give rise to an error of between 2.5 and 10%. Therefore, the quantum efficiency

using the ratio of electrons measured should be from 0.056 to 0.11 root sum square error.

3. *Noise Equivalent Photon Irradiance.* The signal from the array for various levels of known photon irradiance incident on the array should be measured. These signal values along with the noise measurement provide a measure of noise equivalent quanta flux (NEQF). The definition is

$$NEQF = E_p/S/N \qquad (9.49)$$

where the photon irradiance is uniform across the array, E_p is the irradiance on the detector, and S/N is the signal-to-noise ratio.

4. *Linearity.* In a manner similar to the NEQF measurements, the signal out versus the input photon irradiance incidence can be measured and plotted. The array should have about a 1000 : 1 dynamic range where the voltage signal out will be directly proportional to the incident photon irradiance.

5. *Noise.* The noise measurements can be done using the classical correlated double-sampling technique. Figure 9-50 shows the output sampled at times t_i and t_z. The signal is obtained by subtracting the output at t_2 from the output at t_1. (Milton, 1980). Figure 9-51 shows the circuitry where the two sample and hold units are fed into a difference amplifier.

The difference between the two voltage levels at times t_1 and t_2 is the important quantity for determining noise. This difference voltage (ΔV_i) is then digitized and stored in computer memory. A 14-bit analog-to-digital (A/D) converter could be used with this system to provide for a range of 16,428, which is a little more than the dynamic range expected from most detector arrays. The number of samples of voltage difference

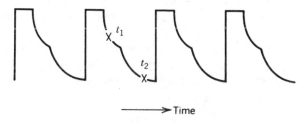

Figure 9-50 Sampling of output from CCD.

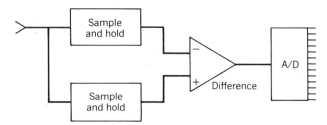

Figure 9-51   Correlate double-sample configuration.

is limited only be the computer memory available. Once the ΔV_i values are in computer memory, the noise can readily be evaluated.

Noise is the standard deviation of several (e.g., 256) output pulses, each for the same no-signal measurement condition

$$\sigma = \left[\frac{\sum_{i=1}^{n} \Delta V_i^2 - \dfrac{1}{n}[\sum_{i=1}^{n} \Delta V_i]^2}{n-1} \right]^{1/2} \tag{9.50}$$

where

$$\Delta V_i = V_i(t_2) - V_i(t_1)$$

6. *Charge Transfer Efficiency.* The charge transfer efficiency should be tested two ways: using a radiation signal input and using an electrical signal input. The reason for the two methods is that the electrical signal-charge transfer is easier to perform and the measurements are relatively straightforward, whereas the measurements with a radiation source are more difficult to perform, but the output data are more meaningful in the evaluation of the array as it is to be used.

The charge transfer efficiency (CTE) evaluation requires putting in a known charge packet at the edge of a row or column and measuring the output after it has transferred through all the gates. The CTE efficiency is usually determined by introducing a given number of signal packets into the charge transfer device, transferring this group down the entire array, and determining the amount of charge lost in the lead pulses. The charge lost in the lead pulses is shifted back to the trailing pulses; but the output of the group may consist of several more pulses than the input because of charge smearing. A first-order approximation to the charge transfer efficiency from Eq. (9-10) is

$$\mathrm{CTE} = (1 - \alpha)^{1/n} \cong 1 - \frac{\alpha}{n} \qquad (\alpha \ll 1) \tag{9.51}$$

where α = fractional charge loss in the leading pulse and n = number of transfers.

The value of each pulse height is read by the computer and the CTE is calculated directly using the above equation. The exact timing for the various readings is critical. It is also important to have a sufficient number of pulses in the group to determine a no-loss pulse height.

In the case of radiation signal measurements of CTE, a four-lens assembly should be provided in front of the array in the environmental test chamber to focus four point sources onto the array. The output from these equal intensity sources would be measured relative to the number of transfers out of the array.

7. *Conversion Efficiency.* The conversion efficiency is found by taking the ratio of the output voltage to the number of electrons coming out of the array. It is given in volts per number of electrons. To measure this important conversion parameter, the test setup shown in Fig. 9-52 should be used in conjunction with the on-chip amplifier.

The electrometer reads the amount of charge in each packet for a given signal voltage (V_0). The conversion efficiency is found by plotting the reset diffusion current I_s in microamperes versus output voltage V_0 in volts as shown in Fig. 9-53.

The slope of this curve gives the conversion efficiency by using the

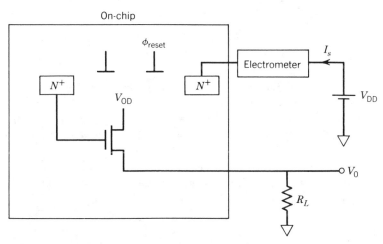

Figure 9-52 Conversion efficiency measurements. V_{OD} = output diffusion; V_{DD} = reset diffusion.

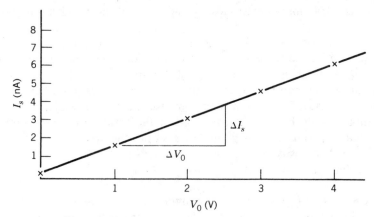

Figure 9-53 Output voltage versus output current.

following equation:

$$V/e = \frac{\Delta V_0}{\Delta I_s} ef \qquad (9.52)$$

where $e = 1.6 \times 10^{-19}$ C,

f = reset frequency.

This information gives a value for the capacitance of the floating diffusion output amplifier. The conversion efficiency relates the voltage produced on a capacitor for a single electron: $V = Q/C$, or for a typical value of 1 pF, one electron will produce $0.16 \ \mu V$.

9-8 TIME-DELAY INTEGRATION (TDI)

For applications in thermal imaging where the signal is a few percent above the background, time-delay integration (TDI) can improve the signal-to-noise ratio. Consider a scanning system using either of the two single-detector elements shown in Figs. 9-54a or 9-54b with a constant scan velocity. Detector a will give us a higher optical resolution but shorter dwell time t_d on the detector, whereas detector b yields poorer optical resolution and longer integration (thus better S/N ratio). Therefore, there is a trade-off between optical resolution and sensitivity (S/N).

By using a time-delay-integration technique with three detectors as shown in Fig. 9-54c, the signal can move over each detector, thus getting the optical resolution of the detector shown in Fig. 9-8a and improving the signal-to-noise ratio. This occurs because signals for this TDI add

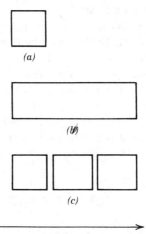

Figure 9-54 Example of time-delay integration: (*a*) single detector (area = A); (*b*) single detector (area = 3A); (*c*) three detectors (area = 3A).

while the noise adds in quadrature (recall Chapter 2) to cause an improvement in SNR of $\sqrt{3}$ in this example. One of the major problems with this technique is that the image scan rate must match the rate of sampling the information from detector to detector. This delay between detectors must be synchronous to the time it takes to scan the image between detectors. This TDI process takes a row of N detectors and makes the signal increase by N and noise by $\sqrt{N}$, producing an improvement in SNR of $\sqrt{N}$ over a single detector. The TDI may be done on or off the focal plane. This synchronism of the scanned image and the summing of

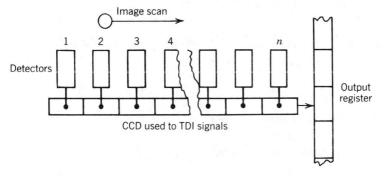

Figure 9-55   CCD-TDI layout.

the signals from each detector is neatly solved by a CCD transfer chain as shown in Fig. 9-55. The CCD is a discrete analog delay line.

The CCD-TDI chain shifts the charge one detector cell to the right in synchronism with the image scan. The target on detector 1 will be on detector N at the Nth time period. The result is that the charge packet transferred onto the output multiplexer has N target signals accumulated in it. The problems associated with TDI are mechanical alignment and CCD well capacity. The CCD well at the Nth pixel is usually filled. Techniques such as skimming the wells have been developed to reduce the DC background charge in the wells to alleviate this problem.

9-9 FIGURES OF MERIT

To give a focal plane a single number to specify its performance is impossible. The entire optical radiation detector business is still struggling to get out from under the photometric nomenclature problem (Biberman, 1967). The figure of merit for a CCD array should take into account at least the filling efficiency of the array, spatial resolution limit, modulation transfer function, quantum efficiency, and integration time (or bandwidth) in order for it to be universally accepted.

If one uses a responsivity expression for array photon responsivity $(V\,q^{-1}\,sec^{-1}\,cm^{-2})$, one can derive an expression for a single pixel. It could then be used as a normalization factor for the MTF response at zero spatial frequency (Barbe, 1975). The responsivity $\mathcal{R}$ of a single pixel can be expressed as

$$\mathcal{R}_V^p = \frac{V_D}{E_p} \qquad (9.53)$$

where V_D = voltage output of the device,
   E_p = uniform photon irradiance on device $(q\,sec^{-1}\,cm^{-2})$.

If monochromatic irradiance is considered, one can rewrite the expression for the output voltage of a pixel as

$$V_D = E_p(\lambda)\Delta x\,\Delta y\,t_{int}\,\eta e(\text{CE}) \qquad (9.54)$$

where $\Delta x\,\Delta y$ = resolution element,
   t_{int} = integration time,
   η = quantum efficiency,
   e = charge,
   CE = conversion efficiency of output preamp (V/e^-).

By substituting V_D in the responsivity expression, the spectral array photon responsivity can then be expressed as

$$\mathscr{R}_p^{\text{array}} = \frac{V_D}{E_P(\lambda)} \equiv \Delta x \, \Delta y \, t_{\text{int}} \eta e(\text{CE}) \tag{9.55}$$

The $\mathscr{R}_p^{\text{array}}$ is for zero spatial frequency. To obtain the resolution effects, one must multiply the responsivity by the MTF of the array in at least two orthogonal directions, thus getting two expressions for spectral array photon responsivity $(\text{V q}^{-1}\,\text{cm}^2\,\text{sec}^{-1}\,\text{mm}^{-1})$ in the x direction and y direction.

To get an expression like noise equivalent power (NEP) for a focal-plane array, one must introduce the noise expression. The noise equivalent quanta flux (NEQF) is defined as

$$\text{NEQF} = \frac{\text{noise}}{\mathscr{R}_P^{\text{array}}} \tag{9.56}$$

where NEQF has the units of $\text{q sec}^{-1}\,\text{cm}^{-2}$ and noise is in V.

The voltage noise of the device is caused by all the contributors as discussed in Section 9-6, or

$$\text{Noise} = \Delta n_n^2 \text{CE} \tag{9.57}$$

where Δn_n^2 is the sum of all the noises added in quadrature.

By using Eq. (9.57) one finds

$$\text{NEQF} = \frac{\sqrt{\Delta n^2}}{\Delta x \, \Delta y t_{\text{int}} \eta} \quad (\text{q sec}^{-1}\,\text{cm}^{-2}) \tag{9.58}$$

This is the photon irradiance flux on the array that would give a signal-to-noise ratio of 1. Again, this noise equivalent photon flux is for spatial frequency zero and the MTF of the array should be normalized to it. This expression indicates how low a photon irradiance level can be detected

In the case of background-limited-infrared-photodetector (BLIP) operation, the noise is associated with the photon irradiance. One can rewrite the NEQF as

$$\text{NEQF} = \frac{\sqrt{2\,G^2 \eta E_p \Delta x \, \Delta y \, \tau_{\text{int}}}}{\Delta x \, \Delta y \, \tau_{\text{int}} \eta}$$

$$\text{NEQF} = G\sqrt{\frac{2E_p}{\Delta x \, \Delta y \, \tau_{\text{int}} \eta}} \tag{9.59}$$

Table 9-8

Available Monolithic Visible (Silicon Response) Imaging Devices (announced by 1983)

Manufacturer	Architecture	Size (V × H)	Pixel mm	Size (μm)	Spacing (μm) x	y	Clock Rate	Noise (e^-/pixel)	Comments	Full Well	Dynamic Range
RCA	Frame transfer	512 × 320	7.3 × 9.75	30 × 30	30	30	TV		Surface channel		
	Frame transfer	512 × 320	7.3 × 9.75					~50–70	Buried, thinned to 10 μm		1000
Fairchild	Interline Transfer	488 × 380	8.8 × 11.4	12 × 18	18	30	TV	~100	Buried channel		
General Electric	CID	244 × 248	0.398 × 0.497	36 × 46	36	46	TV	~250	CID	10^6	
	CID	244 × 380	10.1 × 12.6				TV				
Reticon	MOS switches	32 × 32					2.5 MHz				
	MOS switches	50 × 50					2.5 MHz				
	MOS switches	100 × 100					10 MHz				
	MOS switches	256 × 256									
Bell Northern	Frame transfer	2048 × 96									
Texas Instruments		1024 × 1024					TV	~50	Buttable		
		800 × 800						~20	JPL–Galileo		
		490 × 328					EuroTV	~150	Virtual phase		
		584 × 390									
Ford Aerospace		1024 × 1024					TV		Buried channel		
		256 × 320									
Thompson CSF Corp	Frame transfer	576 × 384	7.25				EuroTV	50	Separated by 1 mm		
		288 × 208									
English Elec Valve GEC	Frame transfer	512 × 320	7.3				TV				
							TV	7			

where G is the photoconductive gain ($G = 1$ for photovoltaic operation, $0 < G < 1$ for photoconductor operation), and τ_{int} is the integration time.

In addition to the single pixel figure of merit of Eq. (9.59), evaluation of the suitability of an array for a specific application would involve examination of imaging system figures of merit. These parameters would include the modulation transfer function as a function of optical radiation wavelength and array fill factor.

9-10 IMAGING ARRAYS

The visible response focal-plane arrays that are presently available (1983) from various manufacturers are shown in Table 9-8. these arrays exhibit a silicon spectral response from the ultraviolet (UV) to 1.1 μm. The table is not complete, but it lists the larger manufacturers in the United States. This section has been included because each manufacturer has created its own type of architecture, and it is interesting to compare the operation of these different architectures.

9-10-1 *RCA*

The focal plane used for visible imaging (silicon response) is a 512×320 pixel array. The transfer of charge is done using a three-phase clocking scheme. The array layout shown in Fig. 9-56 is a frame transfer architecture. The array is compatible with standard television video in that two fields make up a frame (Rodgers, 1974), which is interlaced 2:1. The sequence for obtaining a frame is to have ϕ_{VA2} "on" while the other phases are off so as to obtain an electronic charge distribution proportional to the optical image in the image area mask shown in Fig. 9-56.

After the integration time, the charge is clocked off the image area onto the storage area, using a three-phase clocking scheme. The image area and the storage area on the array are both 256×320 pixels. The image area is now capable of filling the pixels in the second field while the third phase is energized and the others are off. This alternate biasing of phases two and three provides the two image fields for interlacing in the standard television format.

During the integration of this second field, the stored field can be shifted off the array using a horizontal shift register. The output amplifier on the array is a floating diffusion type.

It is interesting to note that if one takes the mask (cover) off the storage area, a full 515×320 array can be utilized. This is in fact what astronomers do when using this array (Aikens, 1982).

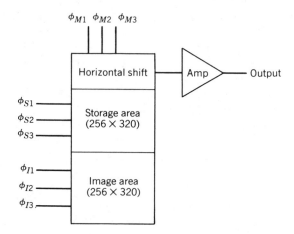

Figure 9-56   RCA's 512×320 array layout.

9-10-2   *Fairchild*

This manufacturer uses an interline transfer approach for a 488×380 pixel array. The CCDs are buried-channel type and are covered with aluminum to prevent the photosensitive CCD shift register from causing image smearing (see Fig. 9-57). The array is TV compatible, in that two fields are interlaced. The clocking of charge is done with two phases. Field one is formed when charge packets from odd-numbered photosites in each column are transferred to the vertical transport registers and read out row by row by the horizontal transport register. Field 2 is read by transferring the charge from the even photosites to the vertical transport register and out. The odd and even fields that come off the device in this manner are interlaced directly.

9-10-3   *Reticon*

This manufacturer developed a new way of using MOSFET switches by using digital signals to access any row. Figure 9-58 shows an equivalent circuit of a two-element array. The line clock causes one voltage pulse to travel down the vertical shift register. At any position in the vertical shift register this pulse activates a horizontal row of MOS switches (Q_1 and Q_2 in Fig. 9-58). The horizontal shift register sequentially pulses columns S_1 through S_2 turning on individual MOSFETs in the row. When a MOSFET is turned on, the photogenerated charge associated with that site is switched directly to the output. Therefore, a single row is transferred off for each vertical shift register voltage pulse.

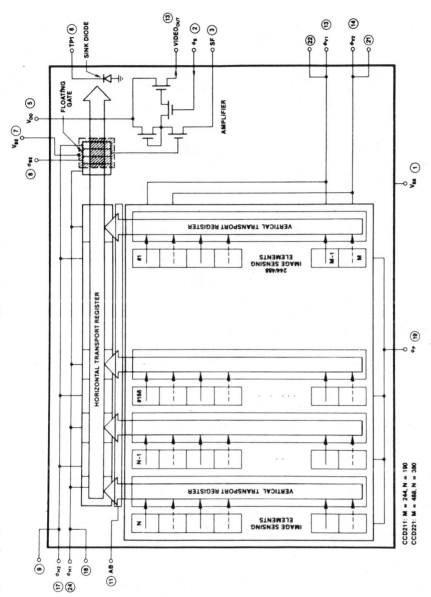

Figure 9-57 Fairchild CCD221 array pin numbers (courtesy of Fairchild).

261

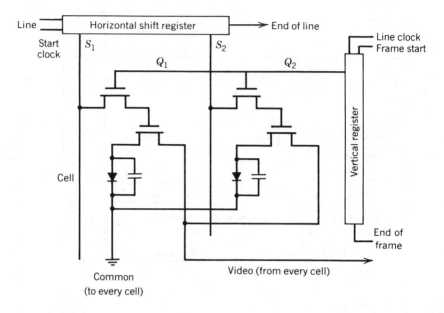

Figure 9-58   Reticon 100×100 array layout showing only two cells (courtesy of Reticon).

9-10-4   *General Electric*

This manufacturer pioneered a different approach to charge transfer called the charge injection device (CID). This architecture provides complete random access to any pixel on the array as well as the ability to read out data nondestructively.

A three-by-three array is shown in Fig. 9-59. Each pixel consists of two MOS capacitors, which are accessible by leads connecting a row (R) or a column (C). Under normal operating conditions, the columns are biased to some voltage commonly (5 V). Now, if a 2 V bias is put on R_1, and C_1 is collapsed to zero volts, the charge for pixel $(1, 1)$ moves under the row electrodes. The charge can be sensed by the row; however, it does not destroy it. The charge can be placed back under the column electrode by raising its voltage again.

Although the array is $x - y$ addressable, it requires more leads from the focal plane and it has greater kTC noise. Since each row has many MOS capacitors on it, the output capacitance for a CID is much larger than for a CCD; therefore, the kTC noise is higher.

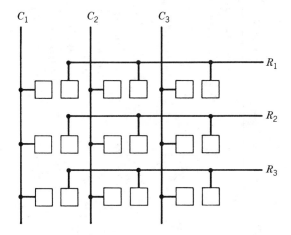

Figure 9-59   CID 3×3 array architecture.

9-10-5   *Texas Instruments (TI)*

The Texas Instruments device, a 490×328 TV-compatible visible array operating in the frame transfer mode, is the result of a new subsurface processing technique known as virtual-phase (VP) technology. This technology improves on the buried-channel technology allowing unidirectional charge transfer with a single clock. The charge transfer mechanism is similar to the "1-1/2" mode employed in two-phase devices; however, the DC level and unidirectionality of the VP device is manufactured into the substrate.

From the user's standpoint, two major advantages were created by this technology:

1. Low dark current ($0.4\,\mathrm{nA\,cm^{-2}}$ at 25°C), as compared to other buried-channel devices ($\sim 10\,\mathrm{nA\,cm^{-2}}$ at 25°C) with no degradation of performance, exhibiting comparable well capacity, filling factor, and good CTE at high speeds (CTE = 0.99997 at 10 MHz)

2. The single clock phase transfer allows for two possible gate formats:

   a.   A patterned gate (no gate over photoactive sites), which furnishes remarkably high QE (particularly in the blue) for a front-face-illuminated array.

   b.   A single gate, providing a uniform broadband optical response. The principle advantage to the producer is the elimination of intergate shorts and the need for overlapping gates to maintain high CTE,

Table 9-9

Infrared-Imaging-Focal-Plane Arrays

Manufacturer	Architecture	Size	Detector Material	Spectral Range (μm)	Operating Temperature (K)	Comment
General Electric	CID	32×32	InSb	1–5.5	77	Random access
RCA	Interline transfer monolithic	64×128	PtSi	1.1–5	77	Schottky barrier
Rockwell	Line-address hybrid	32×32	Si:In	1–8	45	Photoconductor
SBRC	Hybrid	7×64	Si:As	1–25	8	Photoconductor
Aerojet	CID	32×32	Si:Bi	1–19	12	Photoconductor
Electrosystems	Hybrid	16×32	Si:—,	Depends on dopant		Photoconductor
Honeywell	Hybrid	64×64	HgCdTe	1–14	50	Photovoltaic
	Monolithic	32×64	PtSi	1.1–5	77	Schottky barrier
Texas Instruments	Hybrid	64×64	HgCdTe	1–14	77	Photovoltaic
Mitsubishi	Monolithic interline transfer	256×256	PtSi	1–5	77	Schottky barrier
Fujitsu	Monolithic interline transfer	64×64	PtSi	1–5	77	Schottky barrier

which should result in higher yields and, consequently, lower costs—a boon to all concerned.

9-10-6 *Infrared Arrays*

The requirements for cooling infrared arrays increase the cost and complexity of a focal-plane array. In the past, each detector element required a lead to come off the cold array to a room-temperature environment. Now with the advent of CCD technology, an array of detectors having 1000 elements need have only 25 leads which reduces the focal plane assembly heat load.

Presently, the more-advanced infrared-focal-plane arrays employ hybrid construction. The two dominant detector materials used for this hybrid construction are HgCdTe (Roberts, 1983; Rode, 1983) and extrinsic silicon (Pomerrenig, 1983). The two exceptions to the hybrid configuration are the InSb and PtSi monolithic arrays discussed in Table 9-9.

Inasmuch as the coupling of a detector array to a CCD shift register was discussed in the focal-plane-architecture section, no further discussion on that approach needs to be made. The CID approach was briefly discussed under the GE approach to visible-focal-plane layout and therefore will not be reviewed again.

There has been substantial progress in the development of Schottky diode infrared charge-coupled arrays for thermal imaging (Ewing et al., 1982; Capone, Taylor and Kosonocky, 1982; Kosonocky and Elabd, 1983). Since this technology is not like that of the other more conventional infrared sensor studied, a brief description of the physics of operation of a P-type Schottky diode will be discussed as well as the monolithic focal plane. This focal plane is a monolithic array of platinum silicide (PtSi) Schottky photodiodes connected to a CCD readout in an interline transfer architecture. The PtSi photodiodes are fabricated by evaporation of platinum onto silicon followed by an annealing heat treatment (Shepherd, 1983).

The array is back illuminated through the silicon as shown in Fig. 9-60. As a result, the diode photoresponse occurs at wavelengths greater than 1.1 μm, where silicon is transparent. Sensing is by internal photoemission, a two-step process. First, photons are absorbed in the metal, resulting in excitation of electrons above the Fermi energy and corresponding creation of vacancies below the Fermi energy. Subsequently, semiconductor valence electrons tunnel over the potential barrier, ψ_{ms}, between the metal and semiconductor to fill these vacancies. In the case of staring-mode operation, where the Schottky electrode floats at a preset bias, the net effect of photoemission is to increase the number of electrons on the

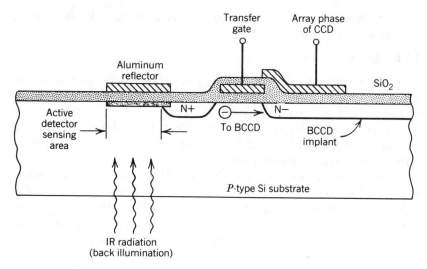

Figure 9-60   The basic Schottky-barrier sensor cell construction (after Ewing et al., 1982).

electrode, thereby reducing its potential. The signal developed at each pixel is proportional to the exposure, the product of the frame time, and the local flux.

The wavelength response of PtSi is to 5 μm; however, other materials, such as iridium (Shepherd, 1981), are being investigated to extend this response further into the infrared.

BIBLIOGRAPHY

Aikens, R. C., private communication, 1982.

Amelio, G. F., "Charge Coupled Devices," *Sci. Am.*, 21, Feb. (1975).

Amelio, G. F., M. F. Tompsett, and G. E. Smith, "Experimental Verification of the Charge Coupled Diode Concept," *Bell Syst. Tech. J.* **49**, 593 (1975).

Barbe, D. F., "Imaging Devices Using the Charge Coupled Concept," *Proc. IEEE*, **63**, 38 (1975).

Barbe, D. F. (ed.), *Charge-coupled Devices*, Springer Verlag, Berlin, (1980).

Barbe, D. F. and S. B. Campana, in *Advances in Image Pickup and Display*, B. Kazan (ed.), Academic Press, New York, 1977, Vol. 3, pp. 171–296.

Biberman, L. M., "Apples, Oranges, and Unlumens," *Appl. Opt.* **6**, 1127 (1967).

Beynon, J. E. and D. L. Lamb, *Charge Coupled Devices and Their Applications*, McGraw-Hill Book Co. (UK) Limited, 1980.

Boyle, W. S. and G. E. Smith, "Charge Coupled Semiconductor Devices," *Bell Syst. Tech. J.* **49**, 587 (1970).

Brodersen, R. W., D. D. Buss, and S. F. Tasch, "Experimental Characterization of Transfer Efficiency in Charge-Coupled Devices," *IEEE Trans. Electron Dev.* **ED-22**, 40 (1975).

Cantella, M., H. Elabd. J. Klein, and W. Kosonocky, "Solid-State focal Plane Arrays Boost IR Sensor Capabilities," *Mil. Electr./Countermeas.* 38–42, Sept. (1982).

Cantella, M. J. "Design and Performance Prediction of Infrared Sensor Systems," *RCA Engineer*, **27**(3), May–June (1982).

Capone, B. R., R. W. Taylor, and W. F. Kosonocky, "Design and Characterization of a Schottky Infrared Charge Coupled Device (IRCCD) Focal Plane Array," *Opt. Eng.* **21**(5), 945 (1982).

Carnes, J. E., W. F. Kosonocky, and E. G. Ramberg, "Drift Aiding Fringing Fields in Charge Coupled Devices," *IEEE J. Solid State Circuits* **SC-6**, 322 (1971).

Carnes, J. E., W. F. Kosonocky, and E. G. Ramberg, "Free Charge Transfer in Charge Coupled Devices," *Trans. IEEE*, **ED-19**, 798, (1972).

Carnes, J. E. and W. F. Kosonocky, "Fast Interface State Losses in Charge Coupled Devices," *Appl. Phys. Lett.* **20**, 261 (1972a).

Carnes, J. E. and W. F. Kosonocky, "Noise Sources in Charge Coupled Devices," *RCA Review* **33**, 327 (1926).

Chan, W. S., "Mosaic Focal Planes: A Technical Rundown," *Optical Spectra*, 57–61, Aug. (1981).

Chapman, R. A. et al., "Monolithic HgCdTe Charge Transfer Device Infrared Imaging Arrays," *IEEE Trans.* **ED-27**, 134–145 (1980).

Considine, P. S., L. A. Lianza, and B. M. Radl, "Image Digitizer System Design Considerations," *Opt. Eng.* **18**, 486–491, (1979).

Deal, B. E., "Standardized Terminology for Oxide Charges Associated with Thermally Oxidized Silicon," *IEEE Trans, Electron Devices* **ED-27**, 606 (1980).

Dereniak, E. L., R. A. Bredthauer, E. M. Hicks, J. E. Vicars, and R. A. Florence, "Microprocessor-Based CCD Test Console," *Proc. SPIE, Staring IR Focal-Plane Tech.*, **267**(9), Feb (1981).

Enders, D. and D. P. Pommerenig, "Correlation between Test Data Obtained at Room and Cryogenic Temperatures for Hybrid Silicon Focal Planes," *SPIE*, **344** (1982).

Esser, L. J. M. and F. L. J. Sangster, in *Handbook on Semiconductors*, C. Hilsum (ed.), North Holland Pub. Co., Amsterdam, 1981, Vol. 4, pp 335–421.

Ewing, W. S., F. D. Shepherd, R. W. Capps, and E. L. Dereniak, "Applications of an Infrared Charge Coupled Device Schottky Diode Array in Astronomical Instrumentation," *SPIE* **331**, 19 (1982).

Fowler, A., P. Waddel, and L. Mortara, "Evaluation of the RCA 512 × 320 CCD Imager for Astronomical Use," *SPIE* **290**(2) (1981).

Hobson, G. S., *Charge Transfer Devices*, Wiley, New York, 1978.

Howes, M. and D. V. Morgan (eds), *Charge-Coupled Devices and Systems*, Wiley Interscience, Chichester, 1979.

Hynecek, J., "Virtual Phase Technology: A New Approach to Fabrication of Large-area CCD's," *IEEE. Trans, on Elect. Dev.* **ED-28**(5), May (1981).

Jensen, W. E., "On Focal Plane CCD Signal Processing Circuits," *SPIE* **217**, 42–54 (1980).

Jordon, A. and N. Jordon, "Theory of Noise in Metal Oxide Semiconductor Devices," *IEEE Trans. Elec. Devices* **ED-12**, Mar. (1965).

Kihata, N., K. Nakumura, E. Saito, and M. Kambra, "The Electronic Still Camera, A New Concept in Photography," *IEEE. Trans.* **CE-28**, 325–331 (1982).

Kim, C. K., *The Physics of CCD and Systems*, Wiley, New York, 1979.

Kimata, M., et al., "PtSi Schottky Barrier IR-CCD Image Sensors," *Jpn. J. Appl. Phys.* **21**, 231–235 (1982).

Kimata, M., M. Denada, N. Yutani, N. Tsubouchi, H. Shibata, H. Kurebayashi, and S. Uematsu, "A 256 × 256 Element Si Monolithic IRCCD Imager," Digest of Technical Papers, International Solid State Conference, pp. 254–255, 1983.

Klein, J. J., N. L. Roberts, and G. D. Chin, "Infrared Camera System Developed in Use the Schottky Barrier IR-CCD Array," *RCA Eng.* **27**(3), May–June, (1982).

Kosonocky, W. F. and J. E. Carnes, "Basic Concepts of Charge Coupled Devices," *RCA Review* **36**, 570 (1975).

Kosonocky W. F., et al., "Schottky-Barrier Infrared Image Sensors," *RCA Eng.* **27**(3), May–June (1982).

Kosonocky, W. F. and H. Elabd, "Schottky-Barrier Infrared Charge Coupled Device Focal Plane Arrays," 27th SPIE International Symposium, Vol. 443, 1983.

Krambeck, R. H., "Zero Charge Transfer," *Bell Syst. Tech. J.* **50**, 10 (1971).

Melen, R. and D. Buss, *Charge-Coupled Devices: Technology and Applications*, IEEE Press, New York, 1977.

Milton, A. F. "Charge Transfer Devices for Infrared Imaging," R. J. Keyes (ed.), in *Optical and Infrared Detectors, Topics in Applied Physics*, Vol. 19, Springer-Verlag, New York, 1980.

Moll, J. L., "Variable Capacitance with Large Capacity Change," Weston Conv. Rec. PT 3, p. 32, 1959.

Nicollian, E. H. and J. R. Brews, *MOS Physics and Technology*, Wiley, New York, 1982.

Ohnishi, K., K. Murakami, and K. Wakui, "Electronic Still-Picture Camera using Magnetic Bubble Memory," *IEEE Trans.* **CE-28**, 321–324 (1982).

Pfan, W. G. and C. G. B. Garrett, "Semiconductor Varactor Using Space-Charge Layers," *IRE* **57**, 2011 (1959).

Pommerrenig, D., "Extrinsic Silicon Focal Plane Arrays," 27th SPIE International Symposium, Vol. 443, 1983.

Roberts, G., "HgCdTe Charge Transfer Device Focal Planes," 27th SPIE International Symposium, Vol. 443, 1983.

Rode, J. P., "Hybrid HgCdTe Arrays," 27th SPIE International Symposium, Vol. 443, 1983.

Rodgers, R. L., "Charge Couple Imager for 525 Line Television," IEEE Intercon, New York, 1974.

Sequin, M. P. and M. F. Tompsett, *Charge Transfer Devices*, Academic Press, New York, 1975.

Shepherd, F. D., "Recent Advances in Schottky IR Photodiodes and Projected Camera Capabilities," IDEM Meeting, Washington, D.C., 1981.

Shepherd, F. D., "Schottky Diode Infrared Detectors," 27th SPIE International Symposium, Vol. 443, 1983.

Shortes, J. R., "Characteristics of Thinned Backside Illuminated Charge Coupled Imagers," *Appl. Phys. Lett.* **24**(11), 565–567 (1974).

Singh, M. P. and D. R. Lamp, "Theoretical Calculations of Surface State Loss in Three Phase Charge Coupled Devices," *J. Phys. D. Appl. Phys.* **9**, 37 (1976).

Stauffer, N. and D. Wilwerding, "Electronic Focus for Cameras," *Scientific Honeyweller*, **3**, 1–13, March (1982).

Sze, S. M., *Physics of Semiconductor Devices*, 2nd ed., Wiley, New York, 1981.

Tanikawa, T., Y. Ito, and A. Shimonashi, "A PtSi Schottky-Barrier Area Imager with Meander-Channel CCD Readout Registers," *IEEE Electron Device Lett.* **EDL-4**, No. 3, 66–67 March 1983.

Terni, Y., et al., "A CCD Imager Using ZnSe-ZnCdTe Heterojunction Photoconductor," *Jpn. J. Appl. Phys.* **21**, Suppl. 21–1, 237–242 (1982).

Thornber, K. K., "Theory of Noise in Charge transfer devices," *Bell Syst. Tech. J.* **53**, 1211 (1974).

Tompsett, M. F., "The Quantitative Effect of Interface States on the Performance of Charge Coupled Devices," *IEEE Trans. Electron devices*, **ED-20**, 45 (1973).

Vicars, J., Rockwell International, private communication, 1982.

Walden, R. H., R. H. Krambeck, R. J. Strain, J. McKenna, N. L. Schryer, and G. E. Smith, "The Buried Channel Charge Coupled Devices," *Bell Syst. Tech. J.* **51**, 1635 (1972).

Younse, J., Texas Instruments, private communication, 1982.

PROBLEMS

9-1.   Describe the "depletion approximation" as related to a MOS capacitor.

9-2.   For a MOS capacitor made on a N-type semiconductor, draw the energy diagrams, the charge distribution for biasing it into accumulation, depletion, and inversion. Discuss the high-frequency limit in terms of mobility of carriers for a N-type versus a P-type semiconductor.

9-3.   For a N-channel device made with 100-μm, square gates, with interelectrode spacing of 125 μm, find the self-induced time constant and thermal diffusion time constant. How much charge loss occurs for a 1-MHz clock due to the above mechanisms?

9-4.   For 512-gate CCD, the loss in the first pulse was 10%, what is the charge transfer efficiency?

9-5.   Discuss the possibility of using channel stops to reduce the interface trapping states.

9-6.   Find an analytic expression for the optimum gate voltage (V_g) applied to a buried-channel CCD.

9-7.   Compare and contrast a surface-channel CCD and a buried-channel CCD.

9-8   What are some of the advantages of a hybrid-focal-plane array over a monolithic array? List some of the disadvantages.

9-9.   Describe how a 2ϕ CCD transfers the charge in a single direction.

9-10.   What is the maximum signal-handling capacity of a silicon MOS capacitor that has the following parameters: Gate Area—0.004 in $\times$0.004 in and Silox thickness = 500 Å. If the system operated at 10 μm what is the maximum $F/\#$ the system could have to be diffraction limited?

9-11.   Explain Fat Zero. What is the edge effect associated with Fat Zero and how does it influence the noise?

9-12.   Why are buried-channel devices faster (higher clocking frequency) than surface-channel devices?

9-13.   What is a typical value for interface surface-state noise (in silicon)?

9-14.   What factors determine charge transfer inefficiency in a CCD?

9-15.   Explain the fill and spill input scheme.

9-16.   What are some of the factors that cause nonlinearities in a CCD focal plane array?

9-17.   Derive the kTC expression using the Johnson noise equivalent circuit.

9-18.   Compare and contrast a CID and a CCD focal-plane array.

9-19.   Explain double-correlated sampling in a CID.

9-20.   Explain conversion efficiency, as applied to a CCD.

9-21.   If you were to build a pyroelectric focal-plane array, how would you approach the design (i.e., monolithic, hybrid)?

9-22.   A CCD has the following characteristics:

$N_e = 100$ (noise electrons)
$N_{sat} = 10^6$ (full well capacity)
$l = 30\ \mu$m   (gate size)
$t_i = 10$ msec   (integration time)
$n = 0.8$ (quantum efficiency)
$\lambda = 1\ \mu$m   (wavelength)

It is mounted behind a camera lens which has the following characteristics:

$D = 50$ mm
$F/\# = 14$
$\tau_0 = 0.5$

Calculate the noise equivalent flux density (NEQF) at 25 lines/mm. (Assume the lens is diffraction limited, the circular entrance pupil is square aperture and the atmospheric transmission is neglected.)

9-23. What is the maximum frequency for a SCCD with a gate length of 10 μm requiring a charge transfer efficiency of 0.999 (Silox thickness 1000 Å)? State any assumptions.

9-24. Derive the expressions for surface potentials for a buried channel CCD for an N-type substrate.

a. Describe how the distance to potential minimum varies with doping of buried channel.

b. What is the value of the bias to ensure only dB CCD operation?

9-25. The surface-interface trap distribution function can be shown to have a factor of 4 in it (compare to Fermi distribution). Explain where this factor came from.

9-26. Derive the CTE expression which is $\text{CTE} \cong 1 - \Delta V / n V_m$. Find where this approximation has a 1% error from the real CTE.

9-27. In what aspects can you model the CCD focal plane as a linear shift invariant system?

9-28. Compare and contrast frame transfer versus interline transfer architecture.

APPENDIX A

PHOTOMETRIC UNITS

Photometric units differ from radiometric units (Chapter 1) in that a standardized human-eye response is embedded in every photometric unit. Photometric units are, therefore, appropriate for measurements of the performance of devices that are to be used with the eye, such as room lights and CRT displays. It is also true that a basic knowledge of photometric units is required to understand the specifications of many devices that have traditionally been measured in photometric units even though the eye is not involved. Low-light-level television cameras are an example from this last group.

Table A-1 lists photometric quantities and units. In order to give meaning to the units, we first define the luminous intensity as a function of the radiance L_e of an area A:

$$I_v = \int_{\lambda=0}^{\infty} \int_A K(\lambda) L_e(\lambda) dA \, d\lambda \qquad (A.1)$$

where $K(\lambda)$ is the luminous efficacy (lumens per watt). The luminous efficacy is related to the human-eye response. The peak value of $K(\lambda)$ for the photopic (bright-light-level adapted) eye response is 673 lumens per watt at $\lambda = 555$ nm. The peak value of luminous efficacy for the scotopic ("dark" adapted) eye response is 1725 lumens per watt at $\lambda = 519$ nm. The normalized eye response

$$V(\lambda) = \frac{K(\lambda)}{K_{\text{peak}}} \qquad (A.2)$$

is shown in Fig. A-1. The definition of the candela is the luminous intensity

Table A-1

Photometric Quantities

Quantity	Symbol	Units		
		Si and mks	cgs	English
Luminous energy	Q_v	Lumen-second (lm s)	Lumen-second (lm s)	Lumen-second (lm s)
Luminous flux	Φ_v	Lumen (lm)	Lumen (lm)	Lumen (lm)
Luminous intensity	I_v	Candela (cd = lm sr^{-1})	Candela (cd = lm sr^{-1})	Candela (cd = lm sr^{-1})
Luminous exitance	M_v	Lux (lx = lm m^{-2})	Phot (ph = lm cm^{-2})	Footcandle (fc = lm ft^{-2})
Illuminance	E_v	Lux (lx = lm m^{-2})	Phot (ph = lm cm^{-2})	Footcandle (fc = lm ft^{-2})
Luminance	L_v	Candela/square meter [cd m^{-2} = nit (nt)]	Stilb (sb = cd cm^{-2})	Candela/square foot (cd ft^{-2})
Luminance of Lambertian surface	L_v	Apostilb (asb = cd π^{-1} m^{-2})	Lambert (L = cd π^{-1} cm^{-2})	Footlambert (fL = cd π^{-1} ft^{-2})

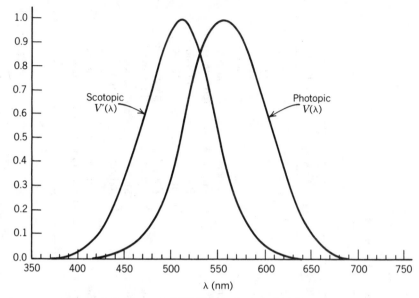

Figure A-1 Scotopic [$V'(\lambda)$] and photopic [$V(\lambda)$] eye response.

of a blackbody with area 1/60th of $1\,\mathrm{cm}^2$ at the temperature of sol-idification of platinum. This yields a reproducible definition for all of the quantities in Table A-1.

Conversion of photometric units to radiometric units can only be performed in those cases for which the spectral distribution of the signal and of the detector response is known. This is a direct consequence of the embedded photopic response in Eq. (A.1).

APPENDIX B

PHOTOVOLTAIC SOLAR CELL PERFORMANCE MODEL

There exist several well-known models for computer calculation of solar cell operation, such as the General Electric model, the Hughes model, the TRW model, and the JPL model (JPL, 1976). These models are more complicated than the simpler model presented here.

The first step in the construction of this model is to obtain an expression for the power produced by a phtotovoltaic device as a function of wavelength. A first approximation to the maximum power that can be produced is to multiply the open-circuit voltage [Eq. (3.14)] by the photocurrent [Eq. (3.12)]:

$$P_{max} = \eta E_e \frac{A_d\lambda}{hc} kT \ln\left[\frac{\eta q E_e A_d\lambda}{hcI_0} + 1\right] \tag{B.1}$$

Extracting the wavelength dependence and normalizing to a numerical value of one at the long-wave length cutoff λ_{max} [Eq. (3.2)]:

$$P_{max}(\lambda) \propto \frac{\lambda \ln \lambda}{\lambda_{max} \ln \lambda_{max}} \tag{B.2}$$

The total power density emitted from a blackbody source of temperature T is

$$\int_0^\infty E_e(\lambda)d\lambda = \int_0^\infty \frac{2\pi hc^2\, d\lambda}{\lambda^5[e^{hc/\lambda kT} - 1]} = \sigma T^4 \tag{B.3}$$

The fraction of this total irradiance that could be transformed into electrical power by a photovoltaic cell with a spectral power response described by proportionality (B.2) is

$$F = \frac{\dfrac{2\pi hc^2}{\lambda_m \ln \lambda_m} \displaystyle\int_0^{\lambda_m} \dfrac{\lambda \ln \lambda \, d\lambda}{\lambda^5 [e^{hc/\lambda kt} - 1]}}{\sigma T^4} \qquad (B.4)$$

Equation (B.4) accounts for losses due to the fact that the solar cell spectral response does not match the spectral emission of the source (see Fig. B-1). The sun is assumed to be a 5800 K blackbody source in this model.

The fraction F from Eq. (B.4) is higher than the efficiencies possible with actual solar cells. A correction must be obtained to account for other power losses. The state-of-the-art models apply individual corrections for junction loss, curve factor losses, recombination, series resistance, reflections, and other factors. The simplified model used here will use only one correction to F to obtain the maximum possible efficiencies as a fraction of F:

$$\eta = KF \qquad (B.5)$$

where K is the proportionality constant to be determined and η is the power conversion efficiency, not the quantum efficiency. Using the least squares criterion, the problem is to find K that minimizes

$$\Delta = \sum_{i=1}^{n} [KF_i - \eta_i]^2 \qquad (B.6)$$

where F_i are determined from Eq. (B.4) for each solar cell for which an

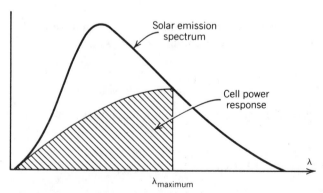

Figure B-1   Solar emission spectrally related to cell response.

actual maximum theoretical efficiency η_i is known. The η_i were taken from p. 561 of Kreith and Kreider (1978). A successive approximation of 100 interations was run on a Radio Shack TRS-80 microcomputer to obtain $K = 0.755$. The final form for the model used is

$$\eta_{max} = \frac{0.755 \times 2 \pi h c^2}{\sigma T^4 \lambda_m \ln \lambda_m} \int_0^{\lambda_m} \frac{\lambda \ln \lambda \, d\lambda}{\lambda^5 [e^{hc/\lambda kT} - 1]} \tag{B.7}$$

Table B-1 lists calculated efficiencies using this model and the efficiencies from Kreith and Kreider (1978) for comparison. The integral in Eq. (B.7) was evaluated using a Riemann sum with $\Delta\lambda = 10$ nm.

Table B-1

Material	e_g (eV)	λ_{max} (μm)	η [Eq. (B.7)]	η [Kreith and Kreider (1978)]
PbSe	0.23	5.38	0.074	
PbS	0.42	2.95	0.140	
Ge	0.67	1.85	0.209	
Si	1.11	1.12	0.260	0.24
InP	1.25	0.99	0.259	0.23
GaAs	1.35	0.92	0.260	0.24
CdTe	1.45	0.85	0.253	0.21
CdSe	1.8	0.69	0.218	
GaP	2.25	0.55	0.160	0.17
CdS	2.4	0.52	0.144	0.16

Now that a model which is usable for feasibility studies has been developed, it can be applied to a specific example: high-efficiency solar-cell-sandwich batteries. Equation (3.2) expresses the fact that photons with wavelength longer than λ_{max} will not interact with the material, but pass right through it. Because the material is nearly transparent to radiant energy longer than λ_{max}, another solar cell with a longer cutoff wavelength than λ_{max} for the top cell could be placed underneath to make use of this formerly wasted energy. A three-layer sandwich might have the spectral power response shown in Fig. B-2. This would result in much higher efficiency than is possible with a single type of cell.

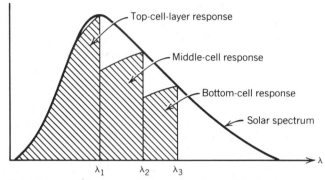

Figure B-2   Solar emission spectrally related to three-cell solar battery response.

The model for a three-cell sandwich is

$$\eta_{max} = \frac{0.755 \times 2\pi h c^2}{\sigma T^4} \left\{ \frac{1}{\lambda_1 \ln \lambda_1} \int_0^{\lambda_1} \frac{\lambda \ln \lambda \, d\lambda}{\lambda^5 [e^{hc/\lambda kT} - 1]} + \frac{1}{\lambda_2 \ln \lambda_2} \right.$$
$$\left. \times \int_{\lambda_1}^{\lambda_2} \frac{\lambda \ln \lambda \, d\lambda}{\lambda^5 [e^{hc/\lambda kT} - 1]} + \frac{1}{\lambda_3 \ln \lambda_3} \int_{\lambda_2}^{\lambda_3} \frac{\lambda \ln \lambda \, d\lambda}{\lambda^5 [e^{hc/\lambda kT} - 1]} \right\} \quad \text{(B.8)}$$

Similar configurations are used for other numbers of sandwich layers. A set of calculated efficiencies using this model is presented in Table B-2.

It is now necessary to include considerations beyond the simple model used to this point. For example, although the last two sandwich cells in Table B-2 have the highest theoretical efficiencies, these cells are not

Table B-2

Sandwich	Theoretical Efficiency
CdS/Si	0.362
CdTe/Si	0.347
CdS/CdTe[a]	0.334
GaP/GaAs[a]	0.346
CdS/CdSe/CdTe[a]	0.367
CdS/Si/Ge	0.441
CdS/Si/Ge/PbS	0.468
CdS/Si/GePbSe	0.462

[a]Best choices; see text.

practical for solar power conversion. This is because the lead salts PbSe and PbS would have to be cooled well below ambient temperature for efficient operation. Inequality (3.3) indicates that the temperature at which a cell will operate is directly proportional to the band-gap energy E_g. Lead selenide and lead sulfide operate most efficiently at a temperature of about $-80°C$ (Eastern Div. Cat). It is desirable, therefore, to obtain high efficiency with large energy gaps to allow high-temperature operation. The three-cell sandwiches footnoted in Table B-2 can all operate at higher temperatures than silicon due to their large band gaps. These three composite cells share another advantage over the other cells shown in the table. All three are made from a single basic material—two from cadmium and one from gallium. This theoretically permits the construction of cell sandwiches by doping a single piece of bulk material with different impurities at different depths. This process would be much less complex than the large-scale integrated circuits, which are produced so cheaply. The theoretical possibility has been shown, therefore, to produce solar photovoltaic sandwich composite cell batteries with an efficiency approaching 37% and higher permissible operating temperatures than silicon. There remains the problem of serious manufacturing difficulties when handling materials of this type. Figure B-3 illustrates the physical configuration of such a device.

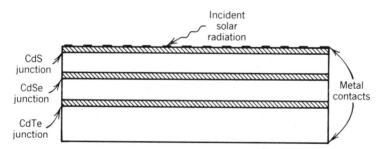

Figure B-3   Cross section of CdS–CdSe–CdTe solar battery.

A simple model of solar cell performance was developed, which is easily programmable on very small computers to yield a design tool for calculating theoretical performance to an accuracy of about 10%. This is sufficient for comparing the theoretical feasibility of new design concepts.

BIBLIOGRAPHY

Infrared Industries, Inc., Eastern Division Catalog.

JPL, *Solar Cell Array Design Handbook*, Volume 1, Section 9.2. NASA, October, 1976.

F. Kreith and J. F. Kreider, *Principles of Solar Engineering*, McGraw-Hill, New York, 1978.

APPENDIX C

BLACKBODY AND DETECTIVITY TABLES

Tabulated values for temperatures of 28 K (liquid hydrogen), 77 K (liquid nitrogen), 197 K (dry ice immersed in acetone), 300 K (room temperature), 500 K, and 1000 K (typical commercial blackbody source temperatures), and 2854 K (tungsten filament incandescent bulb) are presented. Liquid helium temperature (4.2 K) is not included, because the integrated photon emission is so low that from 0 to 20 μm there are only 5.7×10^{-58} photons cm^{-2} sec^{-1} and even from 0 to 50 μm there are only 4×10^{-14} photons cm^{-2} sec^{-1}, which is essentially zero background.

The blackbody cumulative distributions for photons emitted

$$\int_0^{\lambda_p} M_p(T, \lambda) d\lambda = \int_0^{\lambda_p} \frac{1.885 \times 10^{23} \, d\lambda}{\lambda^4 [e^{14,388/\lambda T} - 1]} \text{ photons cm}^{-2} \text{ sec}^{-1} \quad \text{(C.1)}$$

and radiant energy

$$\int_0^{\lambda_p} M_e(T, \lambda) d\lambda = \int_0^{\lambda_p} \frac{3.74151 \times 10^4 \, d\lambda}{\lambda^5 [e^{14,388/\lambda T} - 1]} \text{ W cm}^{-2} \quad \text{(C.2)}$$

are tabulated. These tables can be used for obtaining the desired integral over an interval λ_1 to λ_2 from

$$\int_{\lambda_1}^{\lambda_2} M(T, \lambda) d\lambda = \int_0^{\lambda_2} M(T, \lambda) d\lambda - \int_0^{\lambda_1} M(T, \lambda) d\lambda \quad \text{(C.3)}$$

The detectivity values given are for background-limited photovoltaic detectors operating below the background temperature ($T_d \ll T_{\text{BKD}}$).

The values of D^*_{BLIP} are $\sqrt{2}$ higher than the corresponding photoconductive values. The expressions used to generate these detectivity tables are, for peak spectral D^*

$$D^*_{\text{BLIP}}(\lambda_p, f) = \frac{\lambda_p}{hc} \sqrt{\frac{1}{2 \int_0^{\lambda_p} M_p(300 \text{ K}, \lambda)\,d\lambda}} \qquad (\text{C.4})$$

and, for blackbody D^*

$$D^*_{\text{BLIP}}(T, f) = \left[\frac{\lambda_p}{hc} \sqrt{\frac{1}{2 \int_0^{\lambda_p} M_p(300 \text{ K}, \lambda)\,d\lambda}} \right]\left[\frac{hc}{\lambda_p} \frac{\int_0^{\lambda_p} M_p(T, \lambda)\,d\lambda}{\sigma T^4} \right] \qquad (\text{C.5})$$

Note that the $D^*_{\text{BLIP}}(T, f)$ values always assume a 300 K background.

T = 28 K

λ_P (μm)	$\int_0^{\lambda_P} M_P(T, \lambda)\,d\lambda$ (photons cm^{-2} sec^{-1})	$D^*_{\text{BLIP}}(\lambda_P, f)$ (cm Hz$^{1/2}$ W^{-1})	$\int_0^{\lambda_P} M_e(T, \lambda)\,d\lambda$ (W cm^{-2})
5	3.483E − 026	9.567E + 031	1.396E − 045
6	6.669E − 019	2.624E + 028	2.233E − 038
7	1.012E − 013	7.856E + 025	2.911E − 033
8	7.519E − 010	1.042E + 024	1.895E − 029
9	7.500E − 007	3.711E + 022	1.684E − 026
10	1.840E − 004	2.632E + 021	3.726E − 024
11	1.631E − 002	3.075E + 020	3.009E − 022
12	6.750E − 001	5.216E + 019	1.144E − 020
13	1.556E + 001	1.177E + 019	2.439E − 019
14	2.267E + 002	3.320E + 018	3.307E − 018
15	2.291E + 003	1.119E + 018	3.125E − 017
16	1.720E + 004	4.357E + 017	2.204E − 016
17	1.011E + 005	1.909E + 017	1.223E − 015
18	4.856E + 005	9.224E + 016	5.555E − 015
19	1.966E + 006	4.839E + 016	2.135E − 014
20	6.885E + 006	2.722E + 016	7.120E − 014
21	2.131E + 007	1.624E + 016	2.103E − 013
22	5.928E + 007	1.020E + 016	5.597E − 013
23	1.503E + 008	6.699E + 015	1.360E − 012
24	3.516E + 008	4.571E + 015	3.056E − 012
25	7.659E + 008	3.226E + 015	6.405E − 012

$T = 28$ K ($Continued$)

λ_P (μm)	$\int_0^{\lambda_P} M_P(T, \lambda)\,d\lambda$ (photons cm^{-2} sec^{-1})	$D^*_{\mathrm{BLIP}}(\lambda_P, f)$ (cm Hz$^{1/2}$ W^{-1})	$\int_0^{\lambda_P} M_e(T, \lambda)\,d\lambda$ (W cm^{-2})
30	1.667E + 010	8.297E + 014	1.175E − 010
35	1.443E + 011	3.290E + 014	8.814E − 010
40	7.056E + 011	1.700E + 014	3.816E − 009
45	2.369E + 012	1.044E + 014	1.152E − 008
50	6.127E + 012	7.213E + 013	2.714E − 008
55	1.314E + 013	5.418E + 013	5.356E − 008
60	2.451E + 013	4.327E + 013	9.272E − 008
65	4.114E + 013	3.619E + 013	1.454E − 007
70	6.358E + 013	3.135E + 013	2.113E − 007
75	9.207E + 013	2.791E + 013	2.893E − 007
80	1.265E + 014	2.540E + 013	3.775E − 007
85	1.666E + 014	2.351E + 013	4.740E − 007
90	2.119E + 014	2.206E + 013	5.766E − 007
95	2.616E + 014	2.097E + 013	6.833E − 007
100	3.152E + 014	2.011E + 013	7.924E − 007

$T = 77$ K

λ_P (μm)	$\int_0^{\lambda_P} M_P(T, \lambda)\,d\lambda$ (photons cm^{-2} sec^{-1})	$D^*_{\mathrm{BLIP}}(\lambda_P, f)$ (cm Hz$^{1/2}$ W^{-1})	$\int_0^{\lambda_P} M_e(T, \lambda)\,d\lambda$ (W cm^{-2})
2	6.847E − 021	8.631E + 028	6.870E − 040
3	1.031E − 007	3.337E + 022	6.933E − 027
4	3.392E − 001	2.452E + 019	1.721E − 020
5	2.505E + 003	3.567E + 017	1.023E − 016
6	8.917E + 005	2.269E + 016	3.051E − 014
7	5.664E + 007	3.321E + 015	1.671E − 012
8	1.233E + 009	8.136E + 014	3.201E − 011
9	1.319E + 010	2.798E + 014	3.063E − 010
10	8.612E + 010	1.217E + 014	1.811E − 009
11	3.933E + 011	6.264E + 013	7.563E − 009
12	1.376E + 012	3.654E + 013	2.440E − 008
13	3.924E + 012	2.343E + 013	6.466E − 008
14	9.548E + 012	1.618E + 013	1.470E − 007

T = 77 K (*Continued*)

λ_P (μm)	$\int_0^{\lambda_P} M_P(T, \lambda)\,d\lambda$ (photons cm^{-2} sec^{-1})	$D^*_{\text{BLIP}}(\lambda_P, f)$ (cm Hz$^{1/2}$ W^{-1})	$\int_0^{\lambda_P} M_e(T, \lambda)\,d\lambda$ (W cm^{-2})
15	2.046E + 013	1.184E + 013	2.960E − 007
16	3.960E + 013	9.079E + 012	5.404E − 007
17	7.046E + 013	7.232E + 012	9.110E − 007
18	1.170E + 014	5.943E + 012	1.438E − 006
19	1.932E + 014	5.012E + 012	2.148E − 006
20	2.732E + 014	4.320E + 012	3.063E − 006
21	3.908E + 014	3.793E + 012	4.200E − 006
22	5.392E + 014	3.383E + 012	5.569E − 006
23	7.212E + 014	3.058E + 012	7.174E − 006
24	9.389E + 014	2.797E + 012	9.012E − 006
25	1.194E + 015	2.584E + 012	1.108E − 005
30	3.034E + 015	1.945E + 012	2.430E − 005
35	5.713E + 015	1.654E + 012	4.063E − 005
40	8.968E + 015	1.508E + 012	5.786E − 005
45	1.252E + 016	1.4366E + 012	7.445E − 005
50	1.613E + 016	1.406E + 012	8.957E − 005

T = 197 K

λ_P (μm)	$\int_0^{\lambda_P} M_P(T, \lambda)\,d\lambda$ (photons cm^{-2} sec^{-1})	$D^*_{\text{BLIP}}(\lambda_P, f)$ (cm Hz$^{1/2}$ W^{-1})	$\int_0^{\lambda_P} M_e(T, \lambda)\,d\lambda$ (W cm^{-2})
1	5.067E − 011	5.017E + 023	1.020E − 029
2	9.419E + 004	2.327E + 016	9.618E − 015
3	8.321E + 009	1.174E + 014	5.750E − 010
4	2.115E + 012	9.820E + 012	1.114E − 007
5	5.362E + 013	2.438E + 012	2.294E − 006
6	4.366E + 014	1.025E + 012	1.582E − 005
7	1.872E + 015	5.772E + 011	5.924E − 005
8	5.435E + 015	3.875E + 011	1.528E − 004
9	1.216E + 016	2.914E + 011	3.093E − 004
10	2.278E + 016	2.366E + 011	5.306E − 004
11	3.756E + 016	2.027E + 011	8.095E − 004
12	5.635E + 016	1.805E + 011	1.133E − 003

λ_P (μm)	$\int_0^{\lambda_P} M_P(T, \lambda)\, d\lambda$ (photons cm^{-2} sec^{-1})	$D^*_{\text{BLIP}}(\lambda_P, f)$ (cm Hz$^{1/2}$ W^{-1})	$\int_0^{\lambda_P} M_e(T, \lambda)\, d\lambda$ (W cm^{-2})
13	7.872E + 016	1.655E + 011	1.489E − 003
14	1.041E + 017	1.550E + 011	1.861E − 003
15	1.317E + 017	1.476E + 011	2.240E − 003
16	1.611E + 017	1.424E + 011	2.615E − 003
17	1.914E + 017	1.387E + 011	2.981E − 003
18	2.224E + 017	1.363E + 011	3.332E − 003
19	2.534E + 017	1.348E + 011	3.665E − 003
20	2.842E + 017	1.340E + 011	3.978E − 003
21	3.145E + 017	1.337E + 011	4.272E − 003
22	3.440E + 017	1.339E + 011	4.544E − 003
23	3.726E + 017	1.345E + 011	4.797E − 003
24	4.002E + 017	1.355E + 011	5.030E − 003
25	4.268E + 017	1.367E + 011	5.245E − 003
30	5.426E + 017	1.454E + 011	6.087E − 003

$T = 300$ K

λ_P (μm)	$\int_0^{\lambda_P} M_P(T, \lambda)\, d\lambda$ (photons cm^{-2} sec^{-1})	$D^*_{\text{BLIP}}(\lambda_P, f)$ (cm Hz$^{1/2}$ W^{-1})	$\int_0^{\lambda_P} M_e(T, \lambda)\, d\lambda$ (W cm^{-2})	$D^*_{\text{BLIP}}(T, f)$ (cm Hz$^{1/2}$ W^{-1})
1	6.078E + 000	1.448E + 018	1.233E − 018	3.796E + 001
2	4.113E + 010	3.521E + 013	4.266E − 009	3.123E + 006
3	5.643E + 013	1.426E + 012	3.996E − 006	1.157E + 008
4	1.000E + 015	3.367E + 011	9.800E − 005	6.532E + 008
5	1.320E + 016	1.554E + 011	5.900E − 004	1.769E + 009
6	4.725E + 016	9.857E + 010	1.806E − 003	3.347E + 009
7	1.132E + 017	7.428E + 010	3.812E − 003	5.181E + 009
8	2.125E + 017	6.196E + 010	6.435E − 003	7.098E + 009
9	3.402E + 017	5.510E + 010	9.413E − 003	8.980E + 009
10	4.884E + 017	5.110E + 010	1.251E − 002	1.076E + 010
11	6.491E + 017	4.875E + 010	1.555E − 002	1.240E + 010
12	8.154E + 017	4.745E + 010	1.842E − 002	1.390E + 010
13	9.817E + 017	4.685E + 010	2.106E − 002	1.525E + 010
14	1.144E + 018	4.674E + 010	2.345E − 002	1.647E + 010

T = 300 K (*Continued*)

λ_P (μm)	$\int_0^{\lambda_P} M_P(T, \lambda)\, d\lambda$ (photons cm^{-2} sec^{-1})	$D^*_{\mathrm{BLIP}}(\lambda_P, f)$ (cm Hz$^{1/2}$ W^{-1})	$\int_0^{\lambda_P} M_e(T, \lambda)\, d\lambda$ (W cm^{-2})	$D^*_{\mathrm{BLIP}}(T, f)$ (cm Hz$^{1/2}$ W^{-1})
15	1.300E + 018	4.697E + 010	2.559E − 002	1.756E + 010
16	1.448E + 018	4.748E + 010	2.749E − 002	1.853E + 010
17	1.587E + 018	4.819E + 010	2.916E − 002	1.940E + 010
18	1.717E + 018	4.905E + 010	3.063E − 002	2.017E + 010
19	1.837E + 018	5.005E + 010	3.192E − 002	2.087E + 010
20	1.949E + 018	5.116E + 010	3.306E − 002	2.149E + 010

T = 500 K

λ_P (μm)	$\int_0^{\lambda_P} M_P(T, \lambda)\, d\lambda$ (photons cm^{-2} sec^{-1})	$D^*_{\mathrm{BLIP}}(\lambda_P, f)$ (cm Hz$^{1/2}$ W^{-1})	$\int_0^{\lambda_P} M_e(T, \lambda)\, d\lambda$ (W cm^{-2})	$D^*_{\mathrm{BLIP}}(T, f)$ (cm Hz$^{1/2}$ W^{-1})
1	2.235E + 009	7.554E + 013	4.600E − 010	1.809E + 009
2	1.061E + 015	2.192E + 011	1.136E − 004	1.044E + 010
3	6.113E + 016	4.333E + 010	4.553E − 003	1.624E + 010
4	4.049E + 017	2.245E + 010	2.364E − 002	1.904E + 010
5	1.168E + 018	1.652E + 010	5.711E − 002	2.028E + 010
6	2.261E + 018	1.425E + 010	9.653E − 002	2.076E + 010
7	3.517E + 018	1.333E + 010	1.349E − 001	2.086E + 010
8	4.798E + 018	1.304E + 010	1.689E − 001	2.077E + 010
9	6.020E + 018	1.310E + 010	1.975E − 001	2.059E + 010
10	7.138E + 018	1.337E + 010	2.209E − 001	2.038E + 010

T = 1000 K

λ_P (μm)	$\int_0^{\lambda_P} M_P(T, \lambda)\, d\lambda$ (photons cm^{-2} sec^{-1})	$D^*_{\mathrm{BLIP}}(\lambda_P, f)$ (cm Hz$^{1/2}$ W^{-1})	$\int_0^{\lambda_P} M_e(T, \lambda)\, d\lambda$ (W cm^{-2})	$D^*_{\mathrm{BLIP}}(T, f)$ (cm Hz$^{1/2}$ W^{-1})
0.2	1.924E − 008	5.149E + 021	1.937E − 026	8.995E + 031
0.3	2.251E + 002	7.140E + 016	1.522E − 016	6.920E + 024
0.4	2.069E + 007	3.140E + 014	1.057E − 011	1.774E + 021
0.5	1.788E + 010	1.335E + 013	7.360E − 009	1.186E + 019
0.6	1.523E + 012	1.736E + 012	5.267E − 007	4.088E + 017

T = 1000 K (*Continued*)

λ_P (μm)	$\int_0^{\lambda_P} M_P(T, \lambda)\, d\lambda$ (photons cm^{-2} sec^{-1})	$D^*_{\mathrm{BLIP}}(\lambda_P, f)$ (cm Hz$^{1/2}$ W^{-1})	$\int_0^{\lambda_P} M_e(T, \lambda)\, d\lambda$ (W cm^{-2})	$D^*_{\mathrm{BLIP}}(T, f)$ (cm Hz$^{1/2}$ W^{-1})
0.7	3.489E + 013	4.232E + 011	1.042E − 005	3.613E + 016
0.8	3.536E + 014	1.519E + 011	9.316E − 005	5.768E + 015
0.9	2.090E + 015	7.030E + 010	4.933E − 004	1.369E + 015
1.0	8.489E + 015	3.876E + 010	1.818E − 003	4.294E + 014
1.1	2.631E + 016	2.422E + 010	5.165E − 003	1.651E + 014
1.2	6.667E + 016	1.660E + 010	1.210E − 002	7.406E + 013
1.3	1.448E + 017	1.220E + 010	2.447E − 002	3.740E + 013
1.4	2.791E + 017	9.462E + 009	4.416E − 002	2.074E + 013
1.5	4.891E + 017	7.659E + 009	7.285E − 002	1.240E + 013
1.6	7.936E + 017	6.414E + 009	1.118E − 001	7.887E + 012
1.7	1.209E + 018	5.520E + 009	1.618E − 001	5.278E + 012
1.8	1.750E + 018	4.859E + 009	2.230E − 001	3.685E + 012
1.9	2.425E + 018	4.357E + 009	2.954E − 001	2.668E + 012
2.0	3.239E + 018	3.968E + 009	3.782E − 001	1.992E + 012
3.0	1.809E + 019	2.519E + 009	1.545E + 000	3.003E + 011
4.0	3.639E + 019	2.305E + 009	2.702E + 000	1.128E + 011
5.0	5.711E + 019	2.363E + 009	3.534E + 000	6.198E + 010

T = 2854 K

λ_P (μm)	$\int_0^{\lambda_P} M_P(T, \lambda)\, d\lambda$ (photons cm^{-2} sec^{-1})	$D^*_{\mathrm{BLIP}}(\lambda_P, f)$ (cm Hz$^{1/2}$ W^{-1})	$\int_0^{\lambda_P} M_e(T, \lambda)\, d\lambda$ (W cm^{-2})	$D^*_{\mathrm{BLIP}}(T, f)$ (cm Hz$^{1/2}$ W^{-1})
0.2	1.143E + 013	2.113E + 011	1.183E − 005	8.053E + 050
0.3	2.355E + 016	6.981E + 009	1.662E − 002	1.091E + 037
0.4	9.200E + 017	1.489E + 009	4.985E − 001	1.189E + 030
0.5	7.617E + 018	6.469E + 008	3.384E + 000	7.618E + 025
0.6	2.954E + 019	3.942E + 008	1.121E + 001	1.195E + 023
0.7	7.499E + 019	2.887E + 008	2.503E + 001	1.170E + 021
0.8	1.469E + 020	2.357E + 008	4.404E + 001	3.612E + 019
0.9	2.431E + 020	2.061E + 008	6.650E + 001	2.400E + 018
1.0	3.583E + 020	1.887E + 008	9.056E + 001	2.732E + 017
1.1	4.864E + 020	1.781E + 008	1.148E + 002	4.601E + 016

T = 2854 K (*Continued*)

λ_P (μm)	$\int_0^{\lambda_P} M_P(T, \lambda)\, d\lambda$ (photons cm^{-2} sec^{-1})	$D^*_{\text{BLIP}}(\lambda_P, f)$ (cm Hz$^{1/2}$ W^{-1})	$\int_0^{\lambda_P} M_e(T, \lambda)\, d\lambda$ (W cm^{-2})	$D^*_{\text{BLIP}}(T, f)$ (cm Hz$^{1/2}$ W^{-1})
1.2	6.216E + 020	1.719E + 008	1.381E + 002	1.041E + 016
1.3	7.593E + 020	1.685E + 008	1.599E + 002	2.955E + 015
1.4	8.957E + 020	1.670E + 008	1.800E + 002	1.003E + 015
1.5	1.028E + 021	1.670E + 008	1.981E + 002	3.931E + 014
1.6	1.155E + 021	1.681E + 008	2.144E + 002	1.731E + 014
1.7	1.276E + 021	1.699E + 008	2.288E + 002	8.393E + 013
1.8	1.389E + 021	1.724E + 008	2.417E + 002	4.410E + 013
1.9	1.496E + 021	1.754E + 008	2.530E + 002	2.480E + 013
2.0	1.594E + 021	1.789E + 008	2.631E + 002	1.478E + 013

Index